03-SW-2

DEVELOPMENT OF A PROTOCOL FOR RISK ASSESSMENT OF MICROORGANISMS IN SEPARATE STORMWATER SYSTEMS

by:

Adam W. Olivieri
Eisenberg, Olivieri & Associates, Inc.

Alexandria Boehm
Stanford University

Christopher A. Sommers
Jeffrey A. Soller
Eisenberg Olivieri & Associates, Inc.
Joseph N. Eisenberg
University of California Berkeley

Richard Danielson
BioVir Laboratories, Inc.

2007

WERF

The Water Environment Research Foundation, a not-for-profit organization, funds and manages water quality research for its subscribers through a diverse public-private partnership between municipal utilities, corporations, academia, industry, and the federal government. WERF subscribers include municipal and regional water and wastewater utilities, industrial corporations, environmental engineering firms, and others that share a commitment to cost-effective water quality solutions. WERF is dedicated to advancing science and technology addressing water quality issues as they impact water resources, the atmosphere, the lands, and quality of life.

For more information, contact:
Water Environment Research Foundation
601 Wythe Street
Alexandria, VA 22314-1994
Tel: (703) 684-2470
Fax: (703) 299-0742
www.werf.org
werf@werf.org

This report was co-published by the following organization. For non-Subscriber sales information, contact:

IWA Publishing/ Portland Customer Services
Commerce Way
Colchester CO2 8HP, United Kingdom
Tel: +44 (0)1206 796351
Fax: +44 (0)1206 799331
www.iwapublishing.com
sales@portland-services.com

Library of Congress Catalog Card Number: 2006928032
IWAP ISBN: 978-1-84339-767-0

EOA, Inc.

The research on which this report is based was funded, in part, by the United States Environmental Protection agency through Cooperative Agreement No. CR-827345-01 with the Water Environment Research Foundation (WERF). Unless an EPA logo appears on the cover, this report is a publication of WERF, not EPA. Funds awarded under the Cooperative Agreement cited above were not used for editorial services, reproduction, printing, or distribution.

ACKNOWLEDGMENTS

The research described herein was carried out as a collaborative effort between multiple organizations. The authors and project team thank the Project Subcommittee for their insightful comments, guidance, and general assistance throughout the project.

Report Preparation

Principal Investigators:
Adam W. Olivieri, Dr. P.H., P.E.
Eisenberg, Olivieri and Associates (EOA), Inc.

Alexandria Boehm, Ph.D.
Stanford University

Project Team:
Christopher A. Sommers, M.S.
Eisenberg, Olivieri and Associates (EOA), Inc.

Jeffrey A. Soller, M.S.
Eisenberg, Olivieri and Associates (EOA), Inc.

Joseph N.S. Eisenberg, Ph.D.
University of California at Berkeley

Richard Danielson, Ph.D.
BioVir Laboratories

Robert Spear, Ph.D.
University of California at Berkeley

Robert Cooper, Ph.D.
BioVir Laboratories

Project Subcommittee

Robert Pitt, P.E., Ph.D., DEE, Chair
University of Alabama

Debi Huffman, Ph.D.
University of South Florida

Kristina Mena, MSPH, Ph.D.
University of Texas

Norman E. LeBlanc
Hampton Roads Sanitation District

Rich Field, Ph.D.
U.S. Environmental Protection Agency

Ariamalar Selvakumar
U.S. Environmental Protection Agency

Guang-yu Wang, Ph.D.
Santa Monica Bay Restoration Commission

Peter Moffa, P.E., D.E.E.
Brown & Caldwell

Water Environment Research Foundation Staff

Director of Research:	Daniel M. Woltering, Ph.D.
Program Managers:	Jane M. Casteline
	Christy Bixler

ABSTRACT AND BENEFITS

Abstract:

Recent research has shown that the concentrations of microbial indicator organisms (e.g., fecal coliforms) in stormwater may be quite high. However, studies have not clearly established relationships between the concentrations of indicator organisms and pathogens in stormwater, or between stormwater indicator organism concentrations and illness. Thus, it is difficult to interpret indicator data collected by local public agencies in the context of potential risk to human health. The primary objective of this investigation was to generate guidance to enable more accurate and defensible evaluations of stormwater microorganism data and the associated risks to human health from exposure to pathogens in stormwater. The investigation consisted of three major tasks: reviewing and summarizing relevant published literature, conducting a web-based data questionnaire and developing a science-directed data collection plan. Topics discussed include: 1) waterborne pathogens that pose the greatest risk to human health, 2) concentrations of pathogens and indicator organisms observed in stormwater, 3) defensible relations between indicator organisms and pathogens of public health concern in stormwater, 4) the environmental fate of pathogens and indicator organisms in stormwater, 5) the use of microbial source tracking (MST) techniques to identify sources of fecal contamination, and 6) the effectivness of stormwater microorganism control technologies and associated costs. Additionally, recommended next steps needed to enable defensible evaluations of stormwater microorganism data and the associated risk to human health from exposure to stormwater are presented. Next steps include the development of interim guidance for management prioritization, pathogens enumeration method development, and pilot and nation-wide data collection programs.

Benefits:

- Enables more accurate evaluations of data on microorganisms in stormwater.
- Provides an evaluation of existing guidance levels for bacteria in recreational waters and the appliablility for stormwater.
- Summarizes existing information and data from readily available literature sources to document the following:
    - Waterborne pathogens that pose the greatest risk to human health.
    - Concentrations of pathogens and indicator organisms observed in stormwater.
    - Defensible relations between indicator organisms and pathogens of public health concern in stormwater.
    - Environmental fate of pathogens and indicator organisms in stormwater.
    - Microbial source tracking (MST) techniques used to identify sources.
    - Effectivness and costs of stormwater microorganism control technologies.
- Develops a data collection program to assess the associated risk to human health from exposure to pathogenic microorganisms in stormwater.

Keywords: Stormwater, pathogen, risk assessment, bacteria, viruses, fecal indicators.

TABLE OF CONTENTS

LIST OF TABLES

LIST OF FIGURES

LIST OF ACRONYMS

AFRI	acute febrile respiratory illness
APHA	American Public Health Association
ARA	antibiotic resistance analysis
AWQC	Ambient Water Quality Criteria
BMP	best management practice
CDC	Centers for Disease Control and Prevention
CFU	colony forming unit
CPE	cytopathic effect
CSO	combined sewer overflow
CUP	carbohydrate utilization profiling
dsDNA	double-stranded deoxyribonucleic acid
DUST	Demonstration Urban Runoff Treatment
EMCs	event mean concentrations
F-RNA	f-specific ribonucleic acid
gDNA	genomic deoxyribonucleic acid
gRNA	genomic ribonucleic acid
GPM	gallons per minute
HAV	hepatitis A virus
ICC-PCR	integrated cell culture-PCR
IMS	immunomagnetic separation
L	liter
LD	library-dependent
LI	library-independent
MCH	magnetic capture hybridization
MMWR	Morbidity and Mortality Weekly Report
MPN	most probable number
MRA	microbial risk assessment
mRNA	messenger ribonucleic acid
MSB	male-specific bacteriophages
MST	microbial source tracking
MUG	murinoglobulin
NPDES	National Pollutant Discharge Elimination System
NPS	non-point source
NRC	National Research Council
NSQD	National Stormwater Quality Database
NTAC	National Technical Advisory Committee
PCR	polymerase chain reaction
PFGE	pulsed-field gel electrophoresis
PFU	plaque forming unit
pH	hydrogen potential (acidity/alkalinity)
QA/QC	quality assurance/quality control
qPCR	quantitative polymerase chain reaction
Rep-PCR	repetitive polymerase chain reaction

RT-qPCR	reverse transcription quantitative polymerase chain reaction
SMURRF	Santa Monica Urban Runoff Recycling Facility
SSO	sanitary sewer overflow
U.S. EPA	United States Environmental Protection Agency
U.S. FDA	United States Food and Drug Administration
U.S. PHS	United States Public Health Service
VBNC	viable but non-culturable
WERF	Water Environment Research Foundation

EXECUTIVE SUMMARY

National Pollutant Discharge Elimination System (NPDES) permits issued to local public agencies in the U.S. commonly include monitoring of bacterial indicator organisms in receiving waters and/or stormwater discharges. Public agencies conduct such monitoring as a means of identifying impacts on beneficial uses and evaluating the effectiveness of best management practices (BMPs) designed to reduce stormwater microorganisms in recreational surface waters. Recent research has shown that the concentrations of these indicator organisms in stormwater may be quite high. However, studies have not clearly established relationships between the concentrations of indicator organisms and pathogens in stormwater, or between stormwater indicator organism concentrations and illness. Thus, it is difficult to interpret indicator data collected by local public agencies in the context of potential risk to human health.

The primary objective of this investigation was to generate guidance to enable more accurate and defensible evaluations of stormwater microorganism data and the associated risks to human health from exposure to pathogenic microorganisms (i.e., pathogens) in stormwater. This project was designed to help public agencies begin to understand how to answer questions such as: which pathogens are present in stormwater; what are the potential health risks associated with exposure to these pathogens; what are the sources of pathogens in stormwater; and what can be done to manage the potential health risks?

ES.1 Research Approach

Considering the complexity and importance of the topic, a research approach was developed to begin addressing the issues presented above. The approach consisted of three major tasks: reviewing and summarizing relevant published literature, conducting a web-based data questionnaire and developing a science-directed data collection plan. The approach was guided by the following research questions associated with evaluating and managing risk to human health from exposure to pathogens in stormwater.

1. What waterborne pathogens pose the greatest risk to human health and what are the concentrations of these pathogens and indicator organisms observed in stormwater?
2. Are there defensible relations between indicator organisms and pathogens of public health concern in stormwater?
3. What is the environmental fate of pathogens and indicator organisms in stormwater?
4. What microbial source tracking (MST) techniques are used to identify sources and assess potential public health risk?
5. What are the most effective control technologies used to reduce microorganisms in stormwater and what are the estimated costs?
6. What is the most appropriate assessment method used to quantitatively characterize potential human health risk associated with exposure to pathogens in stormwater?

7. What critical data are needed to enable the assessment of risk associated with exposure to pathogens in stormwater and what is the recommended approach?

ES.2 Pathogens of Primary Health Concern

This portion of the investigation evaluated what waterborne pathogens pose the greatest risk to human health, based on published literature. Although a wide range of pathogens can be transmitted through non-foodborne[1] routes of exposure it has been suggested that relatively few are responsible for the majority of illnesses in the U.S. from known pathogens. Therefore, from a public health perspective, the primary concern is with those few pathogens. The estimated total annual number of illnesses caused by known pathogens in the U.S. is 38.6 million, adjusting for a significant number of unreported illnesses. Of these illnesses, an estimated 63.8% are from non-foodborne transmission pathways, which include water.

A list of pathogens that are likely to be the most important when considering human exposure to stormwater was developed based on:

1. the Center for Disease Control's (CDC) estimated disease burden in the U.S.;
2. the list of pathogens identified by the U.S. EPA during their research related to combined and sanitary sewer overflows; and,
3. pathogens known to be present in wastewater.

These pathogens include: Norwalk-like viruses, Hepatitis A, Astrovirus, and rotavirus (Human Viruses); *Cryptosporidium parvum*, *Giardia lamblia*, *Entamoeba histolytica*, and *Toxoplasma gondii* (Protozoa); and *Campylobacter* spp., *Salmonella* spp., *Shigella* spp., and *E. coli* (0157:H7, enterotoxigenic, and other diarrheogenic) (Bacteria).

ES.3 Microorganisms in Stormwater

Relevant published literature was compiled and summarized to determine what concentrations of pathogens and indicator organisms are observed in stormwater. In parallel, a nationwide data search for gray literature on concentrations of microorganisms in stormwater was conducted via a web-based questionnaire. Based on the information collected, fecal indicator bacteria concentrations in freshwater creeks and conduits of stormwater are higher than those in receiving waters like coastal embayments, fresh water springs, and surf zones. In stormwater, fecal streptococci and *Enterococcus* concentrations are generally lower than those of fecal or total coliform bacteria. A limited number of investigations have monitored for viral indicators in stormwater. Summaries of the published studies reporting fecal indicator and viral indicator data for stormwater and/or storm influenced samples are presented in Table ES-1[2] and Table ES-2, respectively.

---

[1] Non-foodborne pathogens include those transmitted by air, water and person-to-person contact.

[2] Table E-1 includes data from Sanitary Sewer Overflow (SSO) and/or Combined Sewer Overflows (CSO) events that may or may not be directly relevant to this investigation.

Table ES-1. Fecal Indicator Concentrations in Stormwater or Storm Influenced Samples.

Organism	# Studies	Location of Studies	Water Body Types	Range	Median	Units	Notes
Total Coliforms	10	California, Arizona, New Jersey, Mexico, Germany	Rivers, stormwater, creeks, coastal waters,	4 - $2x10^6$	26,250	/100mL	
Fecal Coliforms	20	California, North Carolina, Louisiana, Arizona, New Jersey, Pennsylvania, Italy, Mexico, Israel, Australia Canada	Rivers, stormwater, creeks, coastal waters, drainage canals,	<1 - $8x10^6$	5,588	/100mL	3 studies include SSO/CSO data
E. Coli	10	Louisiana, California, Florida, New Jersey, Canada, Germany	Rivers, stormwater, creeks, coastal waters,	110 - $2x10^6$	9,177	/100mL	1-2 studies include SSO/CSO data
Fecal Streptococcus	4	California, Arizona, New Jersey, Germany	Creeks, stormwater	7.5 - 13,600	3,902	/100mL	1 study includes SSO/CSO data
Enterococcus	7	California, Louisiana, Mexico, Italy, Australia	Rivers, stormwater, wetland	<1 - 80,000	5,001	/100mL	

Table ES-2. Viral Indicator Concentrations in Stormwater and Storm Influenced Samples.

Organism	# Studies	Location of Studies	Water Body Types	Range	Median	Units	Notes
Coliphage	2	California	Rivers	0.5 - 7597	1,983	/100mL	
F^+ Coliphage	2	California	Rivers	<2 - 853	289	/100mL	

Additionally, data was obtained from the National Stormwater Quality Database (NSQD) and the web-based questionnaire. The NSQD includes roughly 3,765 sampling events from 360 sites in 65 communities throughout the U.S. Reported nationwide median concentrations for fecal coliforms, fecal streptococci, total coliforms and *E. coli* were 5091 MPN/100mL, 17,000 MPN/100mL, 12,000 MPN/100mL and 1750 MPN/100mL, respectively. Nearly all agencies and organizations who responded to the web-based questionnaire reported that they had collected fecal indicator data. Respondents indicated that the most widely used indicator was fecal coliform bacteria.

Only a handful of bacterial pathogens have been quantified in stormwater and reported in the published literature. These include *Salmonella* spp., *Pseudomonas aeruginosa*, *Staphylococcus aureus*, *Bacillus* spp. and *Clostridium perfringens*. The following viruses have been reported in stormwater: rotavirus, adenovirus, hepatitis A, and enterovirus. *Giardia* and *Cryptosporidium* are the only protozoan intestinal parasites that have been quantified in stormwater and described in the published or gray literature. A summary of the information compiled from published literature sources on bacterial pathogens, viruses and protozoan parasites is presented in Table ES-3.

Additional pathogen data not presented in Table ES-3 was compiled from respondents to the web-based questionnaire and during the gray literature review. Although a limited number of questionnaire respondent indicated that they had enumerated pathogens in stormwater, one reported *Giardia* concentrations between 10 and 33,000 cysts/100L in stormwater. Additionally, Schroeder et al. (2002) enumerated adenovirus (five samples), enterovirus (one sample), *Salmonella* (1 sample), *Staphylococcus* (2 samples), *Giardia* (1 sample), and *Cryptosporidium* (1 sample) in stormwater.

Table ES-3. Pathogen Concentrations in Stormwater and Storm Influenced Samples.

Type of Data	Organism	# Studies	Location of Studies	Water Body Types	Range	Median	Units	Notes
Bacteria	*Salmonella* spp.	4	Maryland, France, Canada	Rivers, creeks, storm water	0.5 - 425	50	/L	All studies Include SSO/CSO data
	Pseudomomas aeruginosa	4	Maryland, New Jersey, Canada	Creeks, storm water	<10 - 100,000	17,475	/L	1 study includes SSO/CSO data
	Staphylococcus aureus	2	New Jersey	Stormwater	120 - 180,000	56,830	/L	
	Clostridium perfingens	4	New Jersey, Hawaii, Germany, Australia	River, wetland	15 - 4,300	1,121	/L	1 study includes SSO/CSO data
Parasites	*Giardia lamblia*	4	North Carolina, Pennsylvania, New Jersey, Germany	Stream, River, Estuary	0 - 7,500	50	/100L	1 study includes CSO data
	Cryptosporidium	7	North Carolina, Pennsylvania, New Jersey, California, West Virginia, Germany, France	Stream, River, Estuary	1 - 950,000	610	/100L	Highest concentrations are due to rangeland runoff
Viruses	Adenovirus	1	California	rivers	880 - 7,500	NA	genomes/L	PCR
	Hepatitis A	1	California	Rivers	Presence in 67% of samples	NA	NA	Samples taken during storms
	Enteroviruses	5	California, Hawaii, Arizona, North Carolina	Rivers, storm water, estuary	0.01 - 19	0.10	/L	1 study includes CSO data
	Rotavirus	1	Arizona	River	0 - 0.24	NA	/L	

ES.4 Pathogens and Indicator Relationships in Stormwater

Most researchers who correlate human illness to fecal indicator bacteria levels have conducted epidemiological studies in waters receiving point inputs of treated or raw sewage. However, few epidemiological studies have tested the health effects of exposure to waters receiving direct and recent stormwater runoff (See Section ES.7). Thus, there is no explicit documentation of the health effects of stormwater, based on epidemiological studies. An alternative approach is to establish correlations between fecal or viral indicators and pathogens.

In this portion of the investigation applicable published literature was reviewed and summarized with the goal of determining if defensible relationships exist between fecal indicators and pathogens in stormwater and storm influenced surface waters. A total of 14 published studies were reviewed. Based on the literature summarized, there are not widely applicable and defensible relations between pathogens and indicators in stormwater and storm-influenced surface waters. Although several studies did find significant correlations between specific (fecal or viral) indicators, the results from the vast majority of investigations do not support extrapolation of their use to other locations.

ES.5 Environmental Fate of Microorganisms in Stormwater

An important step in the investigation was to evaluate what is known about the environmental fate of microorganisms in stormwater. Microorganisms may enter a waterbody in the free phase or adsorbed to suspended particles. Once in the waterbody, microbes may be preyed upon, become inactivated, adsorb/desorb from particles, or be deposited within sediments at the bottom of the waterbody. In sediments, predation, inactivation and growth rates are more favorable to organism survival than in the water column. When sediments are subjected to high shear stress due to waves, currents, or tides they may be scoured and become resuspended into the water column.

Inactivation results from various mechanisms such as lysis or mutation of genetic material. It is mediated by warm temperatures, sunlight, low nutrients, or high salinity. Non-spore-forming bacteria and viruses are more readily inactivated by these stressors than certain protozoans and spore-forming bacteria. The influence of nutrients and metals on an organism's inactivation/persistence is not straightforward, probably because the nutritional requirements of many bacteria found in stormwater are not completely understood. On the other hand, the effects of temperature and salinity have been examined by numerous researchers. Additionally, an increased association between microorganisms and particles, either suspended or deposited, decreases inactivation rates. In some cases, the growth of bacteria has been observed in sediments. Evidence suggests that particles provide nutrients and shielding from harsh environmental conditions to stormwater microbes. The experimental methods used to determine inactivation rates have not been standardized, so extension of inactivation rates from studied to unstudied systems should be done cautiously. Of special concern are the findings of showing that the nutritional state or phase of a seed pathogen may influence its inactivation rates.

Predation of microorganisms in stormwater has proven to be extremely important in reducing their numbers in waterbodies. Bacteria, protozoa, and viruses have all been shown to be removed, to varying extents in natural waters. The association of microorganisms in stormwater with particles or their deposition in the sediment has been shown to reduce predation pressure.

Up to 100% of fecal bacteria or coliphage have been reported to be bound to suspended sediment in waterbodies. Modelers have taken various approaches to explaining this observation, including assuming adsorption is an equilibrium process and occurs following classical sorption isotherms. The deposition of microbes to sediments can occur whether or not microorganisms are sediment associated. The process of deposition and its role in microorganism fate deserves more study because it is an important link between water column and sediment. It is extremely important to understand the conditions under which microorganisms are deposited in the sediment because they can survive for extended periods and may even be able to reproduce there. Sediment represents an environment where microbes are sheltered from predation and environmental conditions conducive to inactivation. Numerous studies have documented the increased persistence of microbes there. In addition, studies have shown that it is possible to isolate pathogens from sediments when they cannot be isolated from the water column. Thus sediment, particularly after a storm, represents a potentially important environmental reservoir of pathogens. Sediments, and associated microorganisms, can be resuspended when a critical shear stress is experienced at the bed of a waterbody. The required stress to cause sediment or microbe resuspension will depend on properties of the sediment including its size and cohesiveness. In general, a critical shear stress may be exceeded during the passage of waves and during increased stream flow.

It is recommended that future work on the environmental fate of microorganisms in stormwater should focus on the movement of microorganisms between the fluid- and particle-associated phases. In particular, work to further understand particle-microbe interactions in the water column, deposition and resuspension are needed. Additionally, it is recommended that stormwater monitoring programs include sediments as a sampling matrix.

ES.6 Source Tracking Methods For Microorganisms in Stormwater

Microbial source tracking (MST) methods are used to distinguish the source(s) (e.g., wild animals, domesticated animals or humans) of fecal indicator organisms enumerated from waterbodies and/or stormwater discharges. Once a source is identified, appropriate management practices may be put into place to control these inputs. Ideally, MST is based upon the belief that a direct correlation exists between an organism found in the environment and the specific animal source of that specific organism. However, MST has also been used in a broader context to define a group of microorganisms that may have originated with groups or sub-groups of host organisms. A literature review was undertaken to summarize the state of the science for MST. The literature search resulted in the retrieval and summaries of 34 published studies.

Microbial source tracking can be classified into three primary areas of method application: 1) host specific; 2) phenotypic; 3) genotypic. These primary areas of method application can be further broken down by whether they are "library-independent" (LI) or "library-dependent" (LD) in nature. Library-independent methods are based upon detection of the target organism that is directly linked to the specific source. For example, if certain species of *Bacteroides* are found only in humans, then the detection of these types of bacteria would indicate a human source of fecal contamination. Library-dependent methods are based upon characterizing microorganisms that are shared by animals and humans, but can be differentiated by characterizing either a phenotypic trait specific to that host-specific strain, or, by subtle genotypic differences. For example, *E. coli* bacteria are found in most warm-blooded animals; however, it is thought that human strains or subtypes can be differentiated from animal types by comparing specific nucleic acid sequences within their respective DNA.

Each MST technique reviewed has specific advantages and limitations. There is a confused history of the application of the currently employed techniques. Certainly, there is no one way of performing a MST study, as it appears as though the nature of the targeted watershed will dictate the application of the various techniques. It is believed that the future of MST lies in the selection of the appropriate combination of methods. One should not rely solely upon microbial measures, but should try to combine with chemical source tracking and general watershed assessment methods.

ES.7 Treatment Controls for Microorganisms in Stormwater

Various studies have assessed the ability of structural treatment controls (e.g., detention basins, retention ponds, bioretention, infiltration basins, stormwater wetlands) to reduce pollutant concentrations and loadings in stormwater. However, research on the efficiency of these treatment controls, or Best Management Practices (BMPs), has historically concentrated on physical and chemical contaminants such as total suspended solids, nutrients and metals. Studies that have documented the removal of microorganisms in stormwater (i.e., microbial indicators and pathogens) by BMPs are limited and use inconsistent methodologies that make comparisons difficult. Despite these challenges, available information was reviewed and summarized with the goal of determining the BMPs that are most effective at reducing the concentrations of microorganisms in stormwater. Additionally, construction and on-going maintenance and operation costs estimates were reviewed. The stormwater BMPs reviewed include: detention basins, retention ponds, infiltration basins and trenches, vegetated swales, stormwater wetlands, sand filters and advanced treatment and disinfection. Information was collected primarily from published sources, although gray literature was also reviewed where appropriate.

When determining the appropriate stormwater treatment control to use in a specific situation, a variety of factors should be considered, including safety, demonstrated effectiveness, process design fundamentals, construction and on-going costs, potential environmental/adverse effects, and the flow and characteristics of the stormwater that may be treated. However, possibly the most important consideration is the desired effluent quality. Although highly variable, concentrations of microbial indicators in discharges from separate storm sewer systems are generally one to two orders-of-magnitude greater than Ambient Water Quality Criteria (AWQC) for bacteria. Therefore, if it is determined that a specific AWQC for indicator bacteria is the effluent goal of a BMP or series of BMPs, the removal efficiency for microorganisms in stormwater should consistently be 99.9% (3.0 log) to 99.99% (4.0 log).

Removal efficiencies for traditional structure treatment BMPs reviewed had removal efficiencies generally less than 90% (<1.0 log) for microorganisms in stormwater. Specifically, the studies reviewed indicate that retention ponds, detention basins, stormwater wetlands and sand filters had moderate removal efficiencies ranging from 0 to 99.0% (0 log to 2.0 log). Although no studies were found on the efficiencies of infiltration trenches and basins in removing microorganisms in stormwater, the theoretical assumption that these systems are very effective is worth future investigation. Additional data on the removal efficiencies of proprietary controls should also be considered as a research need.

Based on the results from the two stormwater advanced treatment case studies discussed, UV irradiation, in combination with pretreatment suspended sediment removal, should be considered as an effective treatment for the removal of indicator bacteria from dry-weather flows. That said, advanced treatment of stormwater should be approached cautiously. The construction of these facilities was completed due to the identification of high levels of microbial indicators, as opposed to pathogens, in dry weather flows from defined drainage areas. Limited evaluation of potential public health risk associated with these flows was conducted prior to construction of the facilities. These advanced treatment systems can also be expensive to construct (~$9 million). Therefore, it is recommended that a cost-benefit analysis, which includes the consideration of risk, should be conducted prior to the construction of an advanced stormwater treatment facility.

Caution should be used when extrapolating these results of advanced stormwater treatment facilities in Southern California to wet-weather discharges (i.e., stormwater). Due to the high level of suspended solids typically present in stormwater, the efficiency of UV may be reduced due to particle occlusion. Additionally, the design capacity needed to treat stormwater discharges may be prohibitive, from both an engineering and cost perspective.

ES.8 A Review of Microbial Risk Characterization Methods and a Recommended Assessment Approach for Stormwater

This portion of the investigation reviewed methods used to characterize potential human health risk associated with exposure to pathogens in water and summarized published literature potentially relevant to stormwater. The goal of this task was to determine the most appropriate method that can be used to quantitatively characterize potential human health risk associated with exposure to pathogens in stormwater.

ES.8.1 Microbial Risk Characterization Methods and Research

There are two general methods that are used to characterize the potential human health risks associated with exposure to pathogens in water; human health effects (epidemiologic) studies and risk assessments. Epidemiologic studies measure illness directly and associate the outcome measure with illnesses attributable to the exposure of interest. For example, a health effect study may look at a variety of outcomes such as gastrointestinal illness, skin rash, and ear infections that may be caused by a specific exposure to water. Risk assessments, on the other hand, generally estimate risk indirectly based on environmental measurements and numeric modeling.

Environmental health effects studies can use a variety of study designs, including intervention, prospective, cross-sectional and ecologic. Many health effects studies have examined the link between exposure to water and illness, although few published studies explicitly address stormwater exposure. These studies include those that focus on recreational water, drinking water and occupational exposures. The most relevant to stormwater were recreational water studies of marine and freshwaters.

A total of 17 marine water studies were identified. Bacterial indicators of fecal contamination considered in these studies were *E. coli*, fecal coliforms and total coliforms. The strongest trend of increasing relative risk with increasing concentration of indicator was associated with *Enterococcus* (which was the only significant trend based on a linear regression). The three nonfecal indicators used in these studies were *Staphylococcus* spp., *Pseudomonas* spp., and *Aeromonas* spp. Of these, only *Aeromonas* spp. resulted in significant risks. The two indicators of viral contamination studied in marine waters were enterovirus and bacteriophage. An association was only found between illness and enterovirus.

In the 10 studies conducted in freshwaters, *E. coli* was the only bacterial indicator clearly associated with an increase in the relative risk of illness for freshwater exposure. Additionally, the viral indicators, bacteriophage and enterovirus, both showed significant relative risks at all dose levels. The nonfecal bacterial indicators described above showed no trends with illness.

The meta-analysis conducted by Wade et al. (2003) provided a quantitative summary of studies conducted in the past 50 years. The studies taken as a group provide a much stronger weight of evidence towards the risks associated with recreational water swimming than any one study taken alone. The analysis suggests that the U.S. EPA *Enterococcus* rule for marine waters and the *E. coli* rule for freshwaters are protective.

The second method generally used to characterize potential human health risks associated with exposure to pathogens in water is microbial risk assessment (MRA). The MRA method evaluates the likelihood of adverse human health effects that can follow exposure to pathogenic microorganisms or to a medium in which pathogens occur. To the extent possible, the MRA process includes evaluation and consideration of quantitative information; however, it also employs qualitative information as appropriate.

There are generally two types of MRA models, static and dynamic. The fundamental difference between these models is that static models do not account for the unique properties of a dynamic infectious disease process (Table ES-4). Specifically, in static models, the number of individuals assumed susceptible to infection is not time varying, whereas in dynamic models that number is time varying. Static models assume that the population may be categorized into two or three epidemiological states: a susceptible state, an infected, and sometimes a diseased state. Susceptible individuals are exposed to the pathogen of interest. The probability of the exposed individuals moving into the infected/diseased state is governed by the dose of the pathogen to which they are exposed and the infectivity of the pathogen (Figure ES-1). Although a number of environmental sources may expose humans to pathogens, static models typically assume that susceptible individuals are exposed to pathogens from the specific pathway under consideration for the investigation. They therefore do not include the potential interactions and implications of multiple routes of exposure.

Table ES-4. Comparison of Static and Dynamic Risk Assessment Models.

Static risk assessment model	Dynamic risk assessment model
The # of individuals susceptible to infection is time invariant	The # of individuals susceptible to infection is time variant
Direct exposure (environment-to-person)	Direct (environment-to-person) and indirect exposure (person-to-person)
Individual-based risk	Population-based risk
Potential for secondary transmission of infection or disease is typically not considered.	Potential for secondary or person-to-person transmission of infection or disease is considered.
Immunity to infection from microbial agents is typically not considered.	Exposed individuals may not be susceptible to infection or disease because they may already be infected or may be immune from infection due to prior exposure.
Dose-response function is the critical health component.	The dose-response function is important; however, factors specific to the transmission of infectious diseases may also be important.

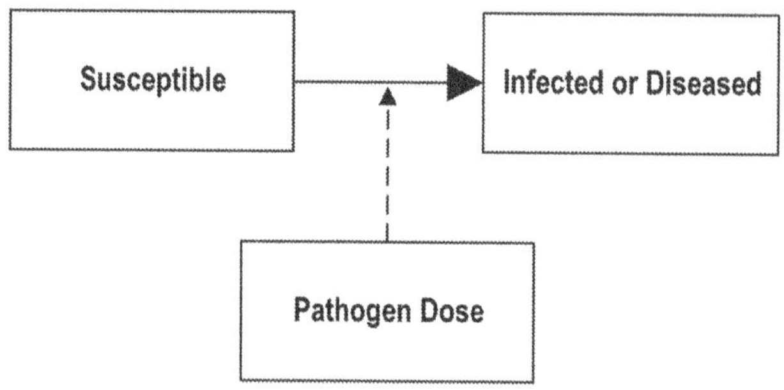

Figure ES-1. Static Risk Assessment Conceptual Model.

A dynamic risk assessment model makes assumptions that divide the population into epidemiological states. Individuals move from one state to another based on epidemiologically relevant data (duration of infection, duration of immunity, etc). Only a portion of the population is in a susceptible state at any point in time, and only those individuals can become infected or diseased through exposure to microorganisms. The probability that a susceptible person moves into an exposed state is governed by the dose of pathogen to which they are exposed; the infectivity of that pathogen; and the number of infected/diseased individuals with whom they may come into contact.

For both dynamic and static representations of the disease process, infectivity as a function of dose (estimated using a dose-response function) is an important factor in estimating risk. The dose-response function is also important in a dynamic microbial risk assessment model, however, other important factors may include person-to-person transmission, immunity, asymptomatic infection, and incubation period. Accounting for these additional factors when estimating risks associated with exposure to pathogenic microorganisms requires a more sophisticated mathematical model than the static model shown conceptually in Figure ES-1. When using a dynamic disease transmission model, one can account for attributes specific to the transmission of infectious diseases. Depending on the infectious disease processes that are important, the dynamic model may include more or fewer components, and therefore vary in complexity. Figure ES-2 shows an example of a dynamic model for reclaimed water, where the population separated into six epidemiological states. Table ES-5 provides a summary of the epidemiological states employed in this dynamic model.

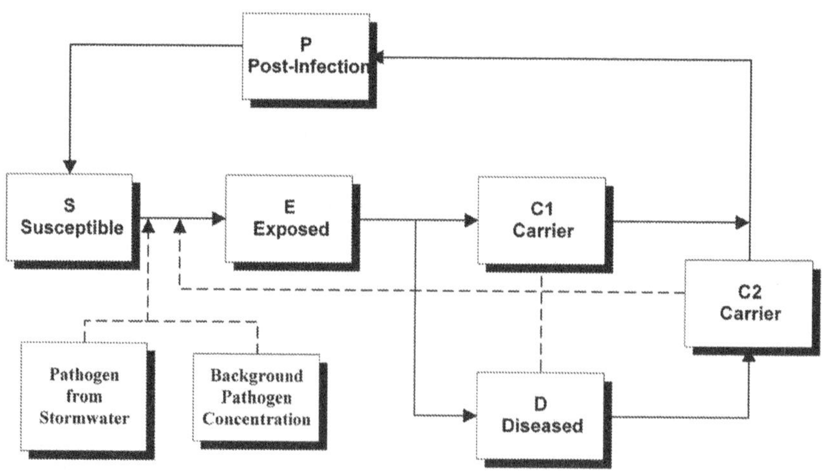

Figure ES-2. Conceptual Model for a Dynamic Risk Assessment.

Table ES-5. Epidemiological States for a Representative Dynamic Model.

Figure ES-2 Label	Epidemiological States	Description
S	Susceptible	Individuals not infected and not protected from infection.
E	Exposed	Individuals exposed to a pathogen, but not yet infected.
C1	Carrier 1	Individuals infected but do not have symptoms of disease.
D	Diseased	Individuals infected and have symptoms of disease.
C2	Carrier 2	Individuals diseased, no longer exhibit symptoms of disease, but still infected.
P	Post-Infected	Individuals neither infected nor symptomatic, and have resistance to infection.

ES.8.2 Recommended Risk Characterization Approach for Stormwater

Few epidemiological studies have specifically addressed the effects of stormwater or non-point sources. Results from these studies have shown the difficulty in drawing conclusive inferences about specific impacts associated with exposures to pathogens in stormwater. Therefore, at this stage, the use of risk assessment models to assess public health concerns associated with stormwater is a valuable alternative to conducting epidemiological studies.

Previous WERF research enables us to evaluate the applicability of MRA techniques for estimating the public health risk from exposure to microorganisms via stormwater. The infectious disease process in a population is, fundamentally, a dynamic process. Therefore, a dynamic model provides the most rigorous and scientifically defensible approach for quantifying the infectious disease process. However, under certain conditions and assumptions, dynamic and static models may yield similar results. These conditions are of particular interest, because modeling the transmission of infectious diseases as a static process requires substantially less data and mathematical sophistication than modeling the process dynamically. In selecting a microbial risk assessment model, however, one must be careful to ensure that any simplifying assumptions are in fact appropriate from an epidemiological perspective.

The fundamental premises upon which the model evaluation results were based are as follows:

- Under the conditions in which the two models predict similar estimations of risk, the static model is more appropriate, as it is simpler yet yields similar results.
- Under the conditions in which the two models predict substantially different estimations of risk, the dynamic model is more appropriate, as one or more infectious disease processes impact the assessment enough to impact the assessment of risk.

To differentiate between the conditions under which the static and dynamic models predict similar and substantially different estimations of risk for the specific exposure scenario under consideration, a series of numerical simulations were set up to explore the range of feasible parameter combinations.

For stormwater, if it is assumed that exposure to microorganisms occurs through recreational activities in receiving waters, the results of the previous WERF research can be used to identify an appropriate risk assessment methodology. The components that need to be considered are the exposure intensity, the dose of pathogens to which individuals would be exposed and the acceptance threshold for difference between the different model results.

The exposure intensity is defined as the product of the proportion of the population exposed and the frequency of exposure (for example, a once per month exposure would have a 1/30 frequency of exposure). For exposures to stormwater it is reasonable to expect that in most cases throughout the U.S., the proportion of the population exposed and the frequency of exposure would both be quite low (for example less than 1/1000th of the population exposed less than once per month), yielding a low exposure intensity.

A limited amount of data is available to characterize pathogen concentrations in stormwater. using assumed volumes of ingestion similar to those assumed for recreational waters and the median concentration values of pathogens in stormwater that are available (Table ES-3), it is possible to estimate on a screening level, exposure doses for stormwater exposures. Based on these assumptions, stormwater exposures could result in virus and parasite exposures that fall in the "medium" or "high" exposure categories, and bacteria exposures that fall into the "high" exposure category.

Based the inferences and assumptions described above it appears that a static MRA methodology may be most appropriate for stormwater risk assessments at this point in time. Once more data are available, it is recommended that the issue of model selection be revisited.

ES.9 Science-Directed Data Collection Program

The principal goal of this investigation is to provide insight relative to the potential human health risk associated with exposure to pathogens in stormwater. Risk associated with these organisms can be quantitatively characterized if data are available to model risk. The literature review conducted in this report revealed that an insufficient number of epidemiology studies relate human illness to stormwater exposure, and that the data available to parameterize human health risk models are not robust. Specifically, there is a striking lack of information regarding the two most important inputs to the static risk model: pathogen concentrations and exposed populations. Therefore, a prioritized, phased approach is presented for acquiring the data necessary to characterize the range of potential human health risks associated with exposure to microorganisms in stormwater. The proposed plan of action emphasizes development of methods and sampling schemes to overcome key inconsistencies in collection of microorganisms in stormwater and analysis that made the few available studies of concentrations of microorganisms in stormwater insufficient for performing a defensible risk assessment.

ES.9.1 Phase 1a: Interim Guidance and Management Prioritization

It is recognized that microorganisms in stormwater currently represent a concern to many municipalities in the U.S. Further, if one applies existing recreational water criteria to non-point source (NPS) dominated waterbodies, the implicit assumption would be that (roughly) the same indicator-to-pathogen relation is present in NPS-impacted water as was present during the epidemiological investigations for recreational waters. Based on the review of the literature, it appears that the available data are insufficient to support that assumption. Thus, a reasonable and appropriate first step in the assessment of risk associated with exposure to microorganisms in stormwater is the development of an interim guidance plan that could provide management prioritization considerations. The emphasis of this recommended interim guidance plan should be the use of available information to facilitate risk management and resource allocation prioritization.

Such interim guidance could be based on qualitative information and expected levels of exposure. For example, it has been suggested that a qualitative approach may facilitate risk management related to recreational water beneficial use impairment. In that case, data from a sanitary survey, relevant water quality data, and specific knowledge about microbiological sources are used along with water quality monitoring and exposure data to derive a relative level of public health concern associated with recreational activities. Thus, management actions could be based on the prioritized level of public health concern.

ES.9.2 Phase 1b: Methods Development

The concentration of pathogens in stormwater is a key input to a risk-based model. There are a handful of methods currently used for determining concentrations of pathogens in stormwater, including nucleic-acid based, culture-based, microscopy-based and hybrid detection techniques. However, few pathogens of public health concern that are likely to be present in stormwater are usually not measurable at levels that can provide insights necessary to make critical management decisions. A static risk model was used to "back calculate" epidemiological significant concentrations of pathogens to determine if methods development/refinement is needed. Estimates indicate that methods should have the goal of detecting greater than or equal to the following: Rotavirus 0.14/L; Non-typhi *Salmonella* 751/L; and, *Cryptosporidium* 1.0 oocyst/L

These detection limits are generally far below those employed by current pathogen detection methods. Therefore, it is recommended that an investment in research to improve techniques for enumerating pathogens in stormwater is a logical first step, in parallel with the development of the interim risk management guidance discussed above. This research should include new approaches to removing inhibitors that are also present in stormwater and may interfere with an enumeration assay.

ES9.3 Phase 2: Pilot Monitoring Program and Risk Assessment in a Single Watershed

Once methods are available to characterize concentrations of pathogens of public health concern at critical levels, a screening-level risk assessment could be conducted. It is recommended that a pilot monitoring program be conducted in a single watershed, followed by a risk assessment investigation. This pilot-scale investigation will encompass many of the aspects of a national monitoring program (land use, sampling location identification, inter- and intra-storm variability, human exposure, etc.), but will represent a modest investment. The lessons learned will increase the likelihood of gathering useful data in a subsequent full-scale investigation.

Methods developed as part of Phase 1b of the phased projects to enumerate important pathogens and indicator organisms in stormwater should be applied to the water and sediment samples collected during the pilot study. As described in Phase 1, it is imperative that the pathogen enumeration procedure contain a step to account for suspended-sediment sorbed pathogens. Each one of the microorganisms identified as target pathogens, as well as a suite of indicators (including *E. coli*, *Enterococcus*, coliphage, and Bacteroides) should be assayed in all stormwater and sediment samples.

Instruments should be deployed to measure both rainfall and volumetric flow rate as functions of time in the stormwater conveyance. Grab samples, which will be used to assay microorganisms, should also be used to assay chemical and physical indicators of water quality including temperature, salinity, nitrate, ammonium, soluble reactive phosphate, turbidity and total suspended solids. Sediment samples should be analyzed for clay, silt and sand content, as well as for organic carbon components.

From the onset of the storm until sampling is complete 2-days after the end of the rain event, the number of individuals recreating in the receiving water and their activities should be recorded. If the area of exposure is small enough, this may be accomplished using a time-lapse video camera. The activity types and exposure frequencies can be used to 1) estimate the volume ingested, and 2) determine the population-based risk of illness using the dynamic risk assessment models.

Using the measured concentrations of pathogen during the storm and the recorded exposure frequency, along with an assumed dilution and die-off factor, one can estimate the risk of illness from exposure to a particular pathogen during a storm event for individuals in a receiving water. These data should be used in conjunction with the risk equations. The range of risks could be compared to risk targets for recreating in freshwater and marine waters (8/1000 and 19/1000, respectively).

The risk distributions from the pilot study can also be used to examine if and how storm characteristics influence median, average, and ranges or risks. Such relationships could subsequently be used to target specific storm types for further studies. These types of relationships could also be used to assign a "concern level" for storms.

ES.9.4Phase 3: National Stormwater Monitoring Program and Risk Assessment

The overall goals of a national stormwater monitoring program will be to quantify 1) the concentration of target infectious/viable pathogens in stormwater at the "end of the pipe" and 2) the number of individuals exposed to stormwater in receiving waters for various types or classes of stormwater.

Once the national stormwater monitoring program is completed, the next goal of this component of the recommended plan of action will be to develop a generalizable risk model. The model should relate risk to easily quantifiable watershed characteristics so that specific watershed types may receive high priority for future study and monitoring. Use of data collected during previous phases in this manner could help direct management actions, inform the potential utility of future epidemiology studies, and inform regulatory limits for microorganisms in stormwater.

CHAPTER 1.0

INTRODUCTION

The primary objective of this investigation was to generate guidance to enable more accurate and defensible evaluations of microorganisms in stormwater[3] and the associated risks to human health from exposure to pathogenic microorganisms (i.e., pathogens) in stormwater. This project was designed to help public agencies begin to understand how to answer questions such as: which pathogens are present in stormwater; what are the potential health risks associated with exposure to these pathogens; what are the sources of pathogens in stormwater; and what can be done to manage the potential health risks?

1.1 Background

National Pollutant Discharge Elimination System (NPDES) issued to local public agencies in the U.S. commonly include monitoring of bacterial indicator organisms in receiving waters and/or stormwater discharges. Public agencies conduct such monitoring as a means of identifying impacts on beneficial uses and evaluating the effectiveness of stormwater best management practices (BMPs) designed to reduce microorganisms in recreational surface waters. Recent research has shown that the concentrations of these indicator organisms in stormwater may be quite high. However, studies have not clearly established relationships between the concentrations of indicator organisms (e.g., fecal and total coliforms) and pathogens in stormwater, or between stormwater indicator organism concentrations and illness. Thus, it is difficult to interpret indicator data collected by local public agencies in the context of potential risk to human health.

State and federal agencies have established guidance levels for indicator organisms in surface waters. However, these guidance levels are based on studies of health effects associated with full-body contact recreation in surface waters impacted with wastewater effluent, as opposed to stormwater. Thus, the applicability of the existing state-adopted ambient water quality criteria for recreational waters warrants critical consideration with respect to stormwater. The fundamental basis for the 1986 Ambient Water Quality Criteria (AWQC) for indicator bacteria was the determination that there is a correlation between the levels of bacterial indicators of fecal contamination in recreational waters to the incidence of acute gastroenteritis infection from recreational water exposures (see Appendix A). This determination was established through epidemiological studies at freshwater and marine beaches that had significant upstream source of sewage contamination (raw, primary, or secondary effluent, both with and without disinfection). Thus, in considering whether the existing guidance levels for microorganisms in recreational waters might be appropriate for stormwater, a number of factors

[3] For the purposes of this report, stormwater is defined as runoff that occurs during or directly after rain or snow events (i.e., wet weather flows). Stormwater does not include water discharged from a storm drainage system that is not associated with a rain or snow event (i.e., dry weather flows).

must be considered including: the potential routes of human exposure to stormwater; the pathogens likely to be present in stormwater; and the relative levels of recreational water indicators and pathogens present in stormwater compared to recreational waters.

Additionally, the temporally transient nature of stormwater will need to be considered carefully if recreational AWQC are used to regulate microorganism concentrations in stormwater. This issue is important because the geometric mean bacterial indicator densities that are used to correspond to accepted gastrointestinal illness rates in recreational waters are derived based on steady state dry weather conditions and are to be based on a statistically sufficient number of samples (generally not less than 5 samples equally spaced over a 30-day period) (U.S. EPA, 1986). The feasibility of meeting these data requirements for evaluating stormwater is questionable, and may need to be evaluated on a case-by case basis. Comparing stormwater concentrations to the single sample maximum values for recreational waters may be another option, but caution will also be required as decisions based on a single sample (or even several samples) may be erroneous (U.S. EPA, 1986).

1.2 Research Approach and Report Organization

Considering the complexity and importance of the topic, WERF and the project team developed a research approach to begin addressing the issues presented above. The approach consisted of three major tasks that are described in this section. The project team's approach was guided by a series of relevant research questions associated with evaluating and managing risk to human health from exposure to pathogens in stormwater. These questions are included in Table 1-1 and revisited throughout the report.

Table 1-1. Primary Research Questions that Directed the Project Team Throughout the Investigation.

1. Which waterborne pathogens pose the greatest risk to human health and what are the concentrations of these pathogens and indicator organisms observed in stormwater?
2. Are there defensible relations between indicator organisms and pathogens of public health concern in stormwater?
3. What is the fate of stormwater microorganisms (i.e., pathogens and indicator organisms) in the environment?
4. What microbial source tracking (MST) techniques are used to identify sources and assess potential public health risk?
5. What are the most effective stormwater microorganism control technologies and what are the estimated costs?
6. What methods have been historically used to quantitatively characterize potential human health risk associated with exposure to waterborne pathogens and what is the most appropriate method for stormwater?
7. What critical data are needed to enable the assessment of risk associated with exposure to pathogens in stormwater and what is the recommended approach?

As a first step, the project team conducted a comprehensive review of available published literature on a variety of topics relevant to assessing the potential public health risk from exposure to pathogens in stormwater. The literature reviews were intended to summarize the most recent information on concentrations of microorganisms in stormwater, fate, source identification, and treatment technologies. The results of the literature reviews are presented in Chapters 2.0-6.0 of this report.

Additionally, WERF understood at the inception of the investigation that a portion of the available data on pathogens and indicators in stormwater might not be present in published literature. Therefore, the literature reviews were augmented with a nationwide questionnaire of WERF subscribers and other public agencies and academic institutions. These data are also summarized in Chapter 2.0.

Data and information collected during the literature reviews and through the questionnaire were then compiled and analyzed to determine whether water quality guidance levels for recreational waters apply to human exposure to pathogens via stormwater. The evaluation considered whether:

- Pathogens in stormwater are likely present in the same relative ratios as those in treated wastewater;
- The routes and magnitudes of exposure for contact with stormwater are different than those that are assumed for full body contact recreation; and,
- Data available to answer the previous two questions in a rigorous and defensible manner.

A discussion of the data evaluation, risk characterization techniques, and a recommendation of the most appropriate method to use when exposure to pathogens in stormwater is of concern is presented in Chapter 6.0.

The final component of the investigation includes a prioritized, phased approach for acquiring the data necessary to characterize the range of potential public health risk associated with the exposure to pathogens in stormwater. The data collection program is designed to provide the information needed to meet the overall objectives of the investigation. The data collection program is presented in Chapter 7.0.

CHAPTER 2.0

PATHOGENS OF PRIMARY HEALTH CONCERN AND MICROORGANISMS IN STORMWATER

1. Which waterborne pathogens pose the greatest risk to human health and what are the concentrations of these pathogens and indicator organisms observed in stormwater?
2. Are there defensible relations between indicator organisms and pathogens of public health concern in stormwater?
3. What is the fate of stormwater microorganisms (i.e., pathogens and indicator organisms) in the environment?
4. What microbial source tracking (MST) techniques are used to identify sources and assess potential public health risk?
5. What are the most effective stormwater microorganism control technologies and what are the estimated costs?
6. What methods have been historically used to quantitatively characterize potential human health risk associated with exposure to waterborne pathogens and what is the most appropriate method for stormwater?
7. What critical data are needed to enable the assessment of risk associated with exposure to pathogens in stormwater and what is the recommended approach?

2.1 Introduction

This chapter provides a brief description of pathogens that can be transmitted through waterborne routes of exposure and assesses which of these pathogens pose the greatest risk to human health. Additionally, the chapter summarizes the results of the published and gray literature review concerning the concentrations of fecal indicators and pathogens in stormwater. The stormwater data presented primarily include fecal indicator and pathogen concentrations in stormwater not directly impacted by combined sewer overflows (CSOs) or sanitary sewer overflows (SSOs). However, in a few of the reported cases, the authors stated, usually in the discussion section of the paper, that CSOs, SSOs, or septic systems could potentially have affected water quality. Lastly, the chapter briefly summarizes information on defensible and widely applicable relations between pathogens and indicators in stormwater and storm influenced surface waters.

2.2 Summary of Waterborne Pathogens of Primary Public Health Concern

Although a wide range of pathogens can be transmitted through waterborne routes of exposure, Mead et al. (1999) suggested that relatively few are responsible for the majority of illnesses in the U.S. from known pathogens. Therefore, from a public health perspective, the primary concern is with those few pathogens.

Mead et al. (1999) estimated the annual total number of illnesses caused by known pathogens in the U.S. to be 38.6 million, adjusting for a significant number of unreported illnesses. Of these illnesses, an estimated 5.2 million cases occur from bacterial pathogens, 2.5 million from parasitic pathogens, and 30.9 million caused by viral pathogens (Table 2-1). Mead and his colleagues also estimated the percentage of these illnesses caused by foodborne pathogens.

Based on these estimates, approximately 90-95% of all non-foodborne4 illnesses in the U.S. are associated with viral enteric pathogens, of which Norwalk-like viruses account for 9.2 million illnesses per year and rotavirus accounts for 3.8 million illnesses per year. The relative importance of viral pathogens in waterborne transmission of disease is supported by data from the (World Health Organization, 1999) and research conducted over the last 20 years on exposure to waterborne pathogens through recreational activities (Cabelli, 1983a; Levine and Stephenson, 1990; Palmateer et al., 1991; Sobsey et al., 1995; Fankhauser et al., 1998).

Table 2-1. Annual Total Disease Burden from Known Pathogens in the U.S.

		Estimated # of Cases	% Non-Foodborne Transmission
BACTERIA			
	Salmonella nontyphoidal	1,400,000	5
	Shigella spp.	450,000	80
	Campylobacter spp.	2,450,000	20
	Others	900,000	
	Bacteria Subtotal	5,200,000	
PROTOZOA			
	Cryptosporidium parvum	300,000	90
	Giardia lamblia	2,000,000	90
	Toxoplasma gondii	230,000	50
	Others	11,000	
	Protozoa Subtotal	2,541,000	
VIRUSES			
	Norwalk	23,000,000	60
	Rotavirus	3,900,000	99
	Astrovirus	3,900,000	99
	Others	83,000	
	Viruses Subtotal	30,883,000	
TOTAL		38,624,000	

(Mead et al., 1999)

Mead et al. (1999) reported that a large percentage of illnesses attributable to protozoans are non-foodborne. As presented in Table 2-1, *Giardia lamblia* causes two million illnesses per year, of which 90% are non-foodborne, and *Cryptosporidium parvum* causes an estimated 300,000 illnesses each year, of which 90% are non-foodborne.

In contrast, many of the important bacterial illnesses in the U.S. are generally associated with a high level of foodborne transmission. Of those bacterial pathogens which account for the non-foodborne illnesses in the U.S., *Salmonella* (*Salmonella* spp. and non-typhoidal), *Shigella*, and *Campylobacter* spp. account for approximately 4.3 million of the 5.2 million annual bacterial

[4] Non-foodborne pathogens include those transmitted by air, water and person-to-person contact.

illness cases in the U.S.(of which approximately 95%, 20%, and 80% are foodborne, respectively).

A list of pathogens likely to be most important when considering human exposure to stormwater include was developed based on:

1. the Center for Disease Control's (CDC) estimated disease burden in the U.S.;
2. the list of pathogens identified by U.S. EPA during their research related to combined and sanitary sewer overflows (U.S. EPA, 2002); and,
3. pathogens known to be present in wastewater.

Pathogens include: Norwalk-like viruses, Hepatitis A, Astrovirus, and rotavirus (Human Viruses); *Cryptosporidium parvum*, *Giardia lamblia*, *Entamoeba histolytica*, and *Toxoplasma gondii* (Protozoa); and *Campylobacter* spp., *Salmonella* spp., *Shigella* spp., and *E. coli* (0157:H7, enterotoxigenic, and other diarrheogenic) (Bacteria).

2.3 Microorganisms in Stormwater

This section summarizes information available in published and gray literature on pathogen and/or fecal indicator concentrations in stormwater, or in receiving waters during wet weather or rain events. Two data collection methods were used to compile this information. First, relevant published literature was collected by searching online services, including PubMed, BIOSIS, and ISI Scisearch. Approximately 600 citation titles and abstracts were reviewed and the most relevant publications were summarized. Information from these publications make up a large portion of this section of the report.

In parallel, the project team also conducted a nationwide data search for gray literature on concentrations of microorganisms in stormwater. The goal of this task was to compile an extensive list of nationwide stormwater pathogen and indicator data to augment data collected during the literature review described above. The project team developed a web-based data request questionnaire and transmittal letter to complete this task. The questionnaire included information on type of available data (i.e., indicator and/or pathogen), type and frequency of sampling, availability of QA/QC data, data format (e.g., hardcopy and/or electronic), and availability of supplemental information regarding human exposure potential. The questionnaire was posted for roughly one month and a total of 63 agencies, organizations and institutions responded. Public agencies and research institutions throughout the U.S. responded. A few responses were also received from international sources. Seven of the 63 respondents indicated that they have attempted to collect and identify pathogens in receiving waters and/or stormwater discharges. The pertinent results from the questionnaire are summarized in this section and all results are included in Appendix B.

2.3.1 Background

Researchers have extensively quantified fecal indicator organisms in stormwater because of their high concentrations and relative ease of enumeration. In contrast, bacterial, viral, and parasitic pathogens often have very low concentrations and are relatively more difficult and time consuming to enumerate. Consequently, there are few studies that have quantified pathogen concentrations, compared to fecal indicators in stormwater,

Summaries of the concentrations of fecal indicators, bacterial pathogens, human viruses, and protozoan pathogens are provided in the following sections, which include tables with information described above. A synthesis section follows in which data sources comparing fecal indicator levels to the pathogen levels or presence/absence are discussed.

2.3.2 Limitations of Available Data

Summaries of data collected via the published and gray literature reviews include information regarding concentrations or presence of microorganisms in stormwater; location and time of study; and detection methods employed. The vast majority of the studies did not document storm information, such as duration and intensity. This may possibly be due to a lack of standard methods to report such data. Some data sources contained concentrations of microorganisms in stormwater for one storm event, while others reported concentrations of microorganisms in stormwater for multiple events over a period of time. Definitions of rainfall events also varied. Interpreting and comparing results from data sources on single storm events with multiple rain events may raise issues of ambiguity. In addition, few of the studies described the sewage infrastructure present at the field site that could potentially affect water quality. This section summarizes storm data and sewage infrastructure information when available. Other deficiencies of information in papers included a lack of detailed descriptions of receiving waters in local communities.

2.3.3 Fecal Indicators

Fecal indicators include fecal indicator bacteria and fecal indicator viruses. Commonly used fecal indicator bacteria include total coliforms, fecal coliforms, *E. coli*, fecal streptococci, and *Enterococcus*. Commonly used fecal indicator viruses include coliphages, F+ coliphages, somatic coliphages, and F-RNA coliphages. Appendix C contains a summary of fecal indicators reported in stormwater, organized by indicator, and reported as number per 100 mL of water. The most common methods used for quantifying fecal bacteria in water were colony forming unit (CFU) methods, including the membrane filtration and the spread plate techniques. In some cases, the most probable number (MPN) method was used to count bacteria concentrations. Plaque forming units (PFU) on specific host lawns were counted to enumerate various phage concentrations.

Total coliforms, fecal coliforms, and *E. coli* concentrations are reported at concentrations in excess of millions of microorganisms in stormwater per 100 mL. The highest concentrations are documented in freshwater creeks and conduits of stormwater. Concentrations are lower in receiving waters like coastal embayments, fresh water springs, and surf zones, where the microorganisms in stormwater have been diluted by cleaner receiving waters. Fecal streptococci and *Enterococcus* concentrations are generally lower than are those of coliform bacteria. The highest reported values of fecal streptococci are on the order of 10,000 MPN/100 mL, with more typical values of 1000 MPN/100 mL. In storm drains in San Diego, CA, *Enterococcus* concentrations on the order of 100 MPN/100 mL are reported (Schiff and Kinney, 2001). Coliphage in stormwater have only been measured by a few researchers (Jiang et al., 2001; Davies et al., 2003; Jiang and Chu, 2004). The reported range is 0-7597 PFU/100 mL.

Additional fecal indicator data were obtained from the National Stormwater Quality Database (NSQD). The NSQD (version 1.1) contains selected water quality information from monitoring carried out as part of the National Pollutant Discharge Elimination System (NPDES) implementation from 1992 to 2002. The NSQD includes roughly 3,765 sampling events from 360 sites in 65 communities from throughout the U.S. Reported nationwide median concentrations for fecal coliforms, fecal streptococci, total coliforms and *E. coli* are 5091 MPN/100mL, 17,0000 MPN/100mL, 12,000 MPN/100mL and 1750 MPN/100mL, respectively (Maestre and Pitt, 2005).

Nearly all agencies and organizations who responded to the web-based questionnaire reported that they had collected fecal indicator data (see Appendix B). Respondents indicated that the most widely used indicator was fecal coliform bacteria. Because of the generally large number of peer-reviewed studies that had concentration data for fecal indicators, these data were not requested from the respondents.

Summaries of the published studies reporting fecal indicator and viral indicator data for stormwater and/or storm influenced samples are presented in Table 2-2 and Table 2-3, respectively. A review of Table 2-2 indicates that the median concentrations of fecal indicators in stormwater are on the order of 103–104 MPN/100mL. The reported concentrations of fecal indicators were highly variable particularly for total coliforms, fecal coliforms, and *E. coli*. It should be noted that several studies summarized in Table 2-2 include data from SSO and/or CSO events and that these events may or may not be directly applicable to this investigation.

For the purposes of this report, viral indicators include coliphages (as a broad class), F+ coliphages, somatic coliphages, and F-RNA coliphages. A limited number of investigations have monitored for viral indicators in stormwater (Table 2-3) and based on the available data, the median concentrations of viral indicators in stormwater are on the order of 102–103 /100mL.

2.3.4 Pathogenic Bacteria

Only a handful of bacterial pathogens have been quantified in stormwater and reported in the peer-reviewed literature. These include *Salmonella* spp., *Pseudomonas aeruginosa*, *Staphylococcus aureus*, *Bacillus* spp. and *Clostridium perfringens*. The latter is used as an indicator and is loosely defined here as a pathogen because it is capable of causing illness in some individuals. The concentrations and other pertinent information about these microorganisms in stormwater, as collected from peer-reviewed studies are documented in Appendix C.

Only one response to the web-based questionnaire indicated that pathogenic bacteria had been enumerated in stormwater. Unfortunately, the data were not available for this report, although numerous attempts were made to retrieve this information.

Salmonella spp. concentrations have been reported in stormwater at sites in the U.S., Europe, and Canada at concentrations between 0.3 and 425 MPN/L (see Appendix C). The highest levels are reported for the European sites (France and the Mediterranean coast).

Pseudomonas aeruginosa has been found in much higher concentrations than *Salmonella* at locations in the U.S. and Canada. Concentrations on the order of 100,000/L, and as low as below the detection limit of the assay, have been reported. In all studies, conduits of stormwater were directly sampled for the pathogen.

Staphylococcus aureus concentrations in stormwater have been quantified in only two studies. Selvakumar and Borst (2006) found that in New Jersey storm drains, concentrations were between 46,000 and 180,000 MPN/L. In a compilation of other studies, O'shea and Field (1992) reported concentrations between 120 and 1200 MPN/L.

Clostridium perfringens has been assayed in U.S. streams, Australian wetlands, and European waterways during wet conditions. The concentrations observed vary between 15 and 4300 CFU/L (see Appendix C). These microorganisms in stormwater are spore formers so they are quite hardy in the environment compared with the non-spore formers already discussed (i.e., *Salmonella*, *Pseudomonas aeruginosa* and *Staphylococcus aureus*)

Fujioka (2001) enumerated *Bacillus* spp. spores in rain-fed streams on Oahu, HI and found the concentrations to be between 88.5 and 1570 CFU/L.

Table 2-2. Fecal Indicator Concentrations in Storm Influenced Samples and Stormwater.

Organism	# Studies	Location of Studies	Water Body Types	Range	Median	Units	Notes
Total Coliforms	10	California, Arizona, New Jersey, Mexico, Germany	Rivers, stormwater, creeks, coastal waters,	4 - $2x10^6$	26,250	/100mL	
Fecal Coliforms	20	California, North Carolina, Louisiana, Arizona, New Jersey, Pennsylvania, Italy, Mexico, Israel, Australia Canada	Rivers, stormwater, creeks, coastal waters, drainage canals,	<1 - $8x10^6$	5,588	/100mL	3 studies include SSO/CSO data
E. coli	10	Louisiana, California, Florida, New Jersey, Canada, Germany	Rivers, stormwater, creeks, coastal waters,	110 - $2x10^6$	9,177	/100mL	1-2 studies include SSO/CSO data
Fecal Streptococcus	4	California, Arizona, New Jersey, Germany	Creeks, stormwater	7.5 - 13,600	3,902	/100mL	1 study includes SSO/CSO data
Enterococcus	7	California, Louisiana, Mexico, Italy, Australia	Rivers, stormwater, wetland	<1 - 80,000	5,001	/100mL	

Table 2-3. Viral Indicator Concentrations in Storm Influenced Samples and Stormwater.

Organism	# Studies	Location of Studies	Water Body Types	Range	Median	Units	Notes
Coliphage	2	California	Rivers	0.5 - 7597	1,983	/100mL	
F^+ Coliphage	2	California	Rivers	<2 - 853	289	/100mL	

2.3.5 Viruses

In the peer-reviewed literature, the presence of the following viruses has been reported in stormwater: rotavirus, adenovirus, hepatitis A, and enterovirus. Appendix C summarizes the concentration or presence of pathogenic viruses in stormwater. All concentrations are reported per liter of water.

Three responses to the web-based questionnaire indicated that viruses had been enumerated in either stormwater or receiving waters. Unfortunately, the data were not available for this report, although numerous attempts were made to retrieve this information.

Rotaviruses have only been quantified by one set of researchers. In an Arizona creek, Rose et al. (1987) found rotavirus in concentrations of up to 0.24 immunofluorescent foci per L . The authors did not document collection during storm events, but mentioned that samples were collected during the recreational seasons between June and August of 1982-1984.

Adenovirus has been measured by Jiang et al. (2001) and Jiang and Chu (2004) in southern California rivers receiving stormwater runoff using PCR[5] nucleic-acid based detection techniques. Using quantitative PCR (qPCR), Jiang et al. (2001) reported 880 to 7500 genomes6 of adenovirus per liter in an urban river. This number is quite high and could be a result of the enumeration method they used (nested PCR). Jiang and Chu (2004) sampled 12 seasonal streams during wet weather and found that eight contained detectable levels of adenovirus (nested PCR presence/absence assay). It should be noted that no assays were conducted by these researchers to show infectivity of the viruses they detected. While RNA tends to be unstable in the environment, it is possible the nucleic double-stranded DNA (dsDNA) from the adenovirus could persist outside the viral particle for an extended period, although this has not been directly tested.

Jiang and Chu (2004) examined the presence of hepatitis A virus in storm-impacted urban rivers and found the RNA from hepatitis A genome in 10 of the 12 samples. It is generally believed that RNA is unstable in the environment, so these observations could be an indication of intact viral particles.

Enterovirus concentrations and presence has been examined by more authors than other viruses in stormwater. Measurements have been made in streams and rivers (Rose et al., 1987; O'Shea and Field, 1992; Lipp et al., 2001; Jiang and Chu, 2004) and in one case a marine bay receiving runoff (Reynolds et al., 1998). Reverse transcriptase, quantitative PCR (RT-qPCR) assays show concentrations between 0 and 0.28 MPN/L. Culture based techniques have shown concentrations up to 19 PFU/L (O'shea and Field, 1992). In a survey of 12 Southern California rivers, 11 were positive for enterovirus based on presence/absence reverse-transcriptase-PCR (RT-PCR) (Jiang and Chu, 2004).

2.3.6 Protozoans

Giardia and *Cryptosporidium* are the only protozoan intestinal parasites that have been quantified in stormwater and described in the peer-reviewed or gray literature. Reported

---

[5] PCR: Polymerase Chain Reaction. A method by which specific DNA sequences are amplified for detection.
[6] qPCR can only detect concentrations of target DNA. In this instance, it is inferred that one copy of the adenovirus gene fragment indicates the presence of an intact dsDNA adenovirus genome. If the detected adenoviruses are infective, then one genome corresponds to one virus.

concentrations are given in Appendix C in units of (oo)cysts per 100 L. The technique utilized to quantify these microorganisms in stormwater consists of microscopy in conjunction with an immunofluorescence antibody stain following a concentration step.

Giardia has been measured in samples of creek and stormwater at levels on the order of 10 cysts/100 L (see Appendix C). In one case, Gibson et al. (1998) found levels as high as 7,500 cysts/100 L, but this sample was likely influenced by a CSO.

Data from one respondent to the web-based questionnaire indicate that *Giardia* concentrations can average between 10 and 33,000 cysts/100L in stormwater (Appendix C). Higher concentrations are generally greater during storm events, as opposed to dry weather flows. It is unclear whether any samples influenced by SSOs.

Reported concentrations for *Cryptosporidium* in the peer-reviewed literature are higher than are those of *Giardia*. In Germany, Kristemann et al. (2002) reported 1.3 to 17.05 oocysts per 100 L in stream water during extreme rain events. Concentrations in rangeland runoff, in the vicinity of cattle fecal deposits are extremely high, 11,250-946,250 oocysts per 100 L (Tate et al., 2000). Other studies reported oocysts in the range of 4 to 15,680 per 100 L.

Xiao et al. (2000) sampled stormwater from brooks near New York City, NY that contribute to the city's drinking water supply. The researchers found that 27 of the 29 50-gallon samples contained *Cryptosporidium* specific DNA (genes for the small subunit rRNA). Microscopy revealed that 14 of 29 samples contained *Cryptosporidium*. The researchers faced challenges using PCR to amplify their target and had to follow up their initial purification with additional purification using IMS (immune-magnetic separation). They concluded that materials in the stormwater significantly inhibited PCR without further purification. Sequences of the amplified product indicated that many of the *Cryptosporidium* spp. isolated were of animal origin. In total, they discovered 12 genotypes among the 27 positive stormwater samples. Only four of the genotypes matched sequences in their library at the time of the study: *Cryptosporidium baileyi*, *Cryptosporidium* from snakes, and *Cryptosporidium* from opossums. Of the remaining eight genotypes, none matched cryptosporidia from humans, farm animals, or companion animals (including *C. felis*, *C. meleagridis*, *C. andersoni*, and *C. parvum*), but nine clustered in the clade containing known human pathogenic *Cryptosporidium* spp. The authors point out the public health consequences of the genotypes found in the storm water samples are not well understood. The study suggests that wildlife is a major source of *Cryptosporidium* in stormwater in New York.

2.3.7 Summary of Pathogens in Stormwater

Table 2-4 presents a summary of the published studies reporting pathogen data for stormwater samples. A review of Table 2-4 indicates that the median concentrations of bacteria are as follows:

- *Salmonella* spp.: 50/L,
- *Pseudomonas aeruginosa*: 1 x 104/L,
- *Staphylococcus aureus*: 1 x 104/L,
- *Clostridium perfringens*: 1 x 103/L, and
- *Bacillus* spp. spores: 800 CFU/L.

Median concentrations of *Giardia lamblia* and *Cryptosporidium parvum* were 50 cysts and 600 oocysts/100L, respectively. The median concentrations of enteroviruses were on the order of 10-1/L. Particular caution is urged in interpreting the data presented in Table 2-4, because some of the studies included samples influenced by CSOs and/or SSOs, and others reflect conditions that may or may not be relevant to urban storm water conditions (e.g., rangeland runoff).

Additional pathogen data not presented in Table 2-4 are described in the following paragraphs. Data were collected by Rajal et al. (2005) at the University of California, Davis (UC Davis) for the California Department of Transportation (CalTrans) in which fecal indicators and pathogen concentrations were monitored in: 1) common areas that produce urban drainage (paved and grass areas of parks, roofs, residential lawns, ponds, storm drains, and other similar surfaces) to develop a broad picture of microbes in urban water (baseline studies); and 2) urban drains and best management practice treatment installations in Southern California (field investigations). The baseline studies were composed of 49 samples taken at 35 sites.

Table 2-4 Pathogen Concentrations in Storm Influenced Samples and Stormwater.

Type of Data	Organism	# Studies	Location of Studies	Water Body Types	Range	Median	Units	Notes
Bacteria	*Salmonella* spp.	4	Maryland, France, Canada	Rivers, creeks, storm water	0.5 - 425	50	/L	All studies Include SSO/CSO data
	Pseudomomas aeruginosa	4	Maryland, New Jersey, Canada	Creeks, storm water	<10 - 100,000	17,475	/L	2 studies include SSO/CSO data
	Staphylococcus aureus	2	New Jersey	Stormwater	120 - 180,000	56,830	/L	Both studies bases on SSO data
	Clostridium perfingens	4	New Jersey, Hawaii, Germany, Australia	River, wetland	15 - 4,300	1,121	/L	1 study includes SSO/CSO data
Parasites	*Giardia lamblia*	4	North Carolina, Pennsylvania, New Jersey, Germany	Stream, River, Estuary	0 - 7,500	50	/100L	1 study includes CSO data
	Cryptosporidium	7	North Carolina, Pennsylvania, New Jersey, California, West Virginia, Germany, France	Stream, River, Estuary	1 - 950,000	610	/100L	Highest concentrations are due to rangeland runoff
Viruses	Adenovirus	1	California	rivers	880 - 7,500	NA	genomes/L	PCR
	Hepatitis A	1	California	Rivers	Presence in 67% of samples	NA	NA	Samples taken during storms
	Enteroviruses	5	California, Hawaii, Arizona, North Carolina	Rivers, storm water, estuary	0.01 - 19	0.10	/L	1 study includes CSO data
	Rotavirus	1	Arizona	River	0 - 0.24	NA	/L	

Schroeder et al. (2002) found significant indicator organism concentrations in most of these samples. Pathogens were detected in 10 of the samples, including adenovirus (five samples), enterovirus (one sample), *Salmonella* spp. (one sample), *Staphylococcus* spp. (2 samples), *Giardia* (1 sample), and *Cryptosporidium* (1 sample). In the samples having the highest fecal indicator counts, no pathogens were detected. Broadly, the results of the field investigations were similar to the baseline studies. Out of 97 samples, one was positive for adenovirus. No other viruses were detected and there were no positives for *Giardia* or

Cryptosporidium. The bacterial detections were limited to enterotoxigenic *E. coli* (4) and *Salmonella* spp. (8). Thus, it was concluded that there does not appear to be a correlation of any kind between the number of indicator organisms and the presence of pathogens.

Following this study, UC Davis continued to establish a quantitative and reliable methodology and QA/QC protocols for 1) the detection of human pathogens in stormwater samples and 2) microbial source tracking to identify the contribution of human versus non-human fecal indicator bacteria to bacterial loads (Rajal et al. 2005). In this study, the researchers collected 56 samples from agricultural, urban and highway locations to validate the filtration method and its combination with quantitative PCR. They detected viral pathogens, the primary targets of this study, in one of 56 samples (human adenovirus was detected once and enterovirus was not detected). In addition, they analyzed sixteen of the fifty-six samples for the presence of other human pathogens, for which qPCR assays had already been designed. In those 16 samples, four were positive for *Cryptosporidium* spp. using an unpublished assay specific for the genus (15-64 copies/100mL), and one was positive for Francisella tularensis (145 copies/100mL).

Microbial source tracking was also performed by UC Davis and bacterial DNA sequences from the Bacteroidales group were detected in 15 of the 17 samples analyzed (Rajal et al. 2005). This indicates that a majority of the samples reflected some non-point sources of fecal contamination. Within the fifteen samples containing Bacteroidales, they detected human marker HF183 Bacteroidales in fourteen samples. The percentage of human to total Bacteroidales ranged from 0.05% to 78.87%.

Lastly, five respondents to the web-based questionnaire indicated that their agency or organization had enumerated *Cryptosporidium* in either stormwater or receiving waters (See Appendix B). One respondent provided data to the project team. *Giardia* concentrations average between 10 and 33,000 cysts/100L in stormwater (Appendix B). Higher concentrations were generally greater during storm events, as opposed to dry weather flows. It is unclear whether any samples were influenced by SSOs.

2.4 Pathogens and Indicator Relationships in Stormwater

Most researchers who correlate human illness to fecal indicator bacteria levels have conducted epidemiological studies in waters receiving point inputs of treated or raw sewage. However, as discussed in Chapter 6.0, few epidemiological studies have tested the health effects of exposure to waters receiving direct and recent stormwater runoff. Thus, there is no explicit documentation of the health effects of stormwater, based on epidemiological studies.

An alternative approach is to establish correlations between fecal or viral indicators and pathogens. In this section, the project team summarizes the applicable literature with the goal of determining if the published literature describes a defensible relationship between fecal indicators and pathogens in stormwater and storm influenced surface waters.

2.4.1 Indicators and Pathogenic Bacteria

Kistenmann et al. (2002) measured *E. coli*, total coliforms, fecal streptococci, C. *perfringens*, *Giardia* and *Cryptosporidium* at three sites in Germany. They took samples in both storm events and non-storm events. They also conducted statistical analyses to evaluate the effect of storms. The researchers found that *E. coli*, total coliforms, fecal streptococci, and C.

perfringens all increase significantly during storm events. However, their correlations were not assessed.

Fujioka (2001) measured *E. coli*, *Enterococcus*, *Clostridium perfringens*, and *Bacillus* spp. spores in rain-fed streams and storm drains on Oahu, Hawaii. He found that in the natural streams, *E. coli* and *Enterococcus* are present in high concentrations; *Clostridium perfringens* in low concentrations and *Bacillus* spp. spores in moderate concentrations. In man-made channels that collect urban runoff and storm runoff, all bacteria were found in high concentrations. He did not perform any explicit correlation analyses.

Baudart et al. (2000) measured *Salmonella* and fecal coliform at sites along the western Mediterranean coast. They found that storm events contribute most of the annual bacteria load. *Salmonella* counts did not correlate with fecal coliforms.

Selvakumar and Borst (2006) reported that no indicator (total coliforms, fecal coliforms, *Enterococcus*, fecal streptococci, *E. coli*) was significantly correlated to *Staphylococcus aureus* in stormwater for urban land uses in New Jersey. *Pseudomonas aeruginosa* was significantly correlated to total coliforms and *E. coli*. However, these two indicators were not correlated to each other.

2.4.2 Indicators and Parasites in Stormwater

Lipp et al. (2001a) measured fecal indicator bacteria (total coliforms, *E. coli*, and *Enterococcus*), *Clostridium perfringens*, enterovirus (via tissue culture), *Cryptosporidium* and *Giardia* (via microscopy) in Charlotte Harbor, an estuary in North Carolina, in both water and sediments. This area has both a municipal sewer system and septic systems. They used logistic regression to relate C. *perfringens*, total coliforms, *E. coli*, *Enterococcus* and coliphage to the occurrence of enteric pathogens. The results indicate that none of the indicators predicted *Giardia* or *Cryptosporidium* concentrations. However, the researchers found that *Enterococcus*, fecal coliform and coliphage correlated significantly with the presence of enteroviruses. Despite the co-occurrence of these organisms in stormwater, Lipp and colleagues detected enteroviruses on days when the water quality standards for the fecal indicator bacteria were not violated. The authors used the data from the whole year and not just from the rainy season. However, they did note that rainfall correlated strongly with fecal indicator organisms and the occurrence of enterovirus in sediments and the water column. The authors only isolated the protozoa at locations in the harbor that had relatively high fecal indicator organism levels.

Gibson et al. (1998) measured *Cryptosporidium* and *Giardia* as well as fecal coliforms from an urban stream in Pittsburgh, Pennsylvania. They knew that CSOs affected the stream during heavy rain events. The researchers could not find a relationship between protozoan parasites and fecal coliforms.

In a study conducted in agricultural lands in West Virginia, Boyer and Kuczynska (2003) found no significant correlation between fecal coliforms (membrane filtration) and *Cryptosporidium* (immunofluorescence). In addition, the authors were not able to relate *Cryptosporidium* to storm flow. One potential explanation proposed for the lack of correlation between fecal indicator and parasites in this study may be the steady inflow of fecal coliforms shed by all warm blooded animals, in contrast to the unsteady inflow of *Cryptosporidium*, which are only shed from infected animals. This may result in variable and unpredictable concentrations of pathogens in the waterbody investigated.

Lemarchand and Lebaron (2003) assessed the relationship between fecal indicators and *Salmonella* and *Cryptosporidium* in a French coastal watershed. They analyzed water samples from nine wastewater treatment plants and eight rivers during the entire year, including the rainy season. They found no clear correlation between the two enteric pathogens or between a given pathogen and fecal indicators. The authors concluded that fecal indicators do not adequately indicate the presence of *Cryptosporidium* or *Salmonella* in natural waters and that pathogens and indicators may have different behaviors in the aquatic environment.

Three drinking water reservoirs were investigated in a project of microbial loading during extreme rainfall and runoff in Germany (Kristemann et al., 2002). The main tributaries were the measuring points for all three locations. One tributary had very little anthropogenic influence, another tributary received heavy runoff from a dairy farm, and the third received treated wastewater discharge. Based on measurements during rainfall and non-rainfall events, this study also concludes that fecal indicators do not show a clear relationship with parasites.

Atherholt et al. (1998) measured concentrations of *Giardia* and *Cryptosporidium* (both via microscopy) in the Delaware River in 1996 to try to relate the concentrations to other microorganisms in stormwater as well as water quality and meteorological parameters. They used the Spearman rank method as the metric for evaluating correlation. *Giardia* correlated significantly with *Cryptosporidium*, total coliforms, *E. coli*, *Clostridium perfringens*, and Coliphage (*Enterococcus* #13706 host). *Cryptosporidium* correlated significantly with *Giardia*, total coliforms, fecal coliforms, *E. coli*, and *Clostridium perfringens*.

2.4.3 Indicators and Viruses

Jiang et al. (2001) measured the concentration of human adenovirus (via nested-PCR), coliphages and male-specific coliphages (via plaque assay) at twelve freshwater outlets in Southern California. The presence of human adenovirus did not correlate significantly with the concentration of coliphages ($r2 = 0.32$) but did significantly correlate with the concentration of male-specific coliphages ($r2 = 0.99$). It was also found that bacterial indicators (total coliforms, fecal coliforms, and *Enterococcus*) did not correlate with the presence of human adenoviruses. Adenoviruses were detected at some sites where the concentration of bacterial indicators did not exceed the state limit, and at other sites adenoviruses were not detected even though the concentration of bacterial indicators exceeded the state limit. Although the paper provided no rainfall data, monitoring occurred during Southern California's wet season.

Jiang and Chu (2004) detected the presence of enterovirus, hepatitis A, and adenovirus at watershed outlets in southern California during storm events. They also measured fecal indicator bacteria and coliphages. The researchers found no relationship between the indicators and the presence of human viruses. In fact, the sites that ranked worst in water quality based on the levels of fecal indicators had no viruses detected, while the site that ranked second best using fecal indicators as a metric had all three viruses present.

Rose et al. (1987) performed viral assays on rotaviruses and enteroviruses, and bacteriological assays on total coliforms, fecal coliforms, and fecal streptococci at Oak Creek Canyon, a freshwater recreational area in Arizona. Seventy-eight percent of their samples that tested positive for enteric viruses met the Arizona coliform recreational water standard for full body contact. The authors found that the correlation coefficient was not significant between fecal coliform and enterovirus or between fecal streptococci and enterovirus. The samples were collected between June and August of 1982-1984, when Arizona usually experiences heavy and frequent storm events.

Olivieri et al. (1977) conducted assays on enterovirus, *Salmonella* spp., P. *aeruginosa*, *Staphylococcus aureus*, total coliforms, fecal coliforms, fecal streptococci, and *Enterococcus*. The concentration of these microorganisms revealed no significant correlation between enteroviruses and bacterial indicators (coliform, streptococci, and *Enterococcus*). In fact, high concentrations of enteroviruses often accompanied low concentrations of bacterial indicators. *Salmonella* spp., P. *aeruginosa*, and *Staphylococcus aureus* showed a somewhat higher correlation with bacterial indicators.

2.4.4 Summary

Based on the data summarized above, the literature does not support widely applicable and defensible relations between pathogens and indicators in stormwater and storm influenced surface waters. Although several studies did find significant correlations between specific (fecal or viral) indicators, the results from the vast majority of investigations do not support extrapolation of their use to other locations.

CHAPTER 3.0

ENVIRONMENTAL FATE OF MICROORGANISMS IN STORMWATER

1. Which waterborne pathogens pose the greatest risk to human health and what are the concentrations of these pathogens and indicator organisms observed in stormwater?
2. Are there defensible relations between indicator organisms and pathogens of public health concern in stormwater?
3. What is the fate of stormwater microorganisms (i.e., pathogens and indicator organisms) in the environment?
4. What microbial source tracking (MST) techniques are used to identify sources and assess potential public health risk?
5. What are the most effective stormwater microorganism control technologies and what are the estimated costs?
6. What methods have been historically used to quantitatively characterize potential human health risk associated with exposure to waterborne pathogens and what is the most appropriate method for stormwater?
7. What critical data are needed to enable the assessment of risk associated with exposure to pathogens in stormwater and what is the recommended approach?

3.1 Introduction

Physical, chemical, and biological constituents in stormwater may differentially affect the fate of pathogens and indicator organisms. Certain microorganisms in stormwater may present a greater threat to human health because of their hardiness in the environment. Association of microorganisms in stormwater with the sediment may create an environmental reservoir that acts as a long-term source of microorganisms to waters. Cost-effective, conservative monitoring programs require knowledge of where and when microorganisms in stormwater reach their most elevated level. The objective of this chapter is to review the possible fates of microorganisms in stormwater in the environment.

Figure 3-1 illustrates possible fates of microorganisms in stormwater in a water-body. Microorganisms in stormwater enter a waterbody either as "free" or "particle-bond" microbes. Within a "parcel" of water/sediment (delineated by the vertical lines in Figure 3-1), a number of processes may occur that influence the fate of a microbe. Inactivation and growth of both free and particle-associated microbes may occur. Predation of both free and particle-associated microbes by eukaryotes or prokaryotes may also occur (in this chapter, we define attack by phage as a form of predation). Free microbes may become adsorbed to (associate with) suspended particles, or particle-associated microbes may become desorbed from (separate from) particles. Microbes may sink to the bottom of the water column where they are deposited and become incorporated into the sediment.

Within the sediment, microbes may experience the same fate processes as in the water column, but the processes may occur at different rates. Finally, turbulence and shear stress at the sediment boundary may cause resuspension of both free and particle-associated microbes.

The conceptual framework illustrated in Figure 3-1 is a mass-balance approach often used in engineering. The processes illustrated represent sources or sinks of microbes in water and sediment. Issues of advection and dispersion in this literature review have not been considered.

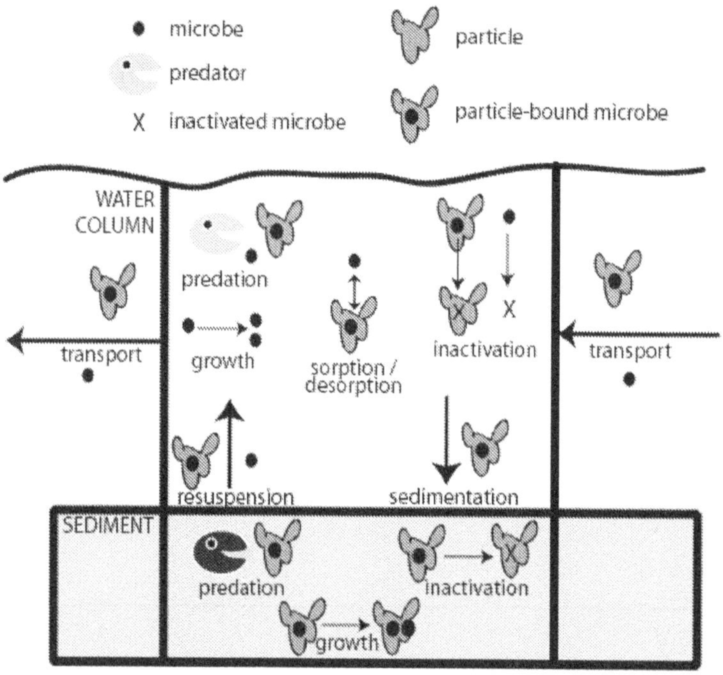

Figure 3-1. The Possible Fates of Microorganisms in Stormwater in Environmental Water and Sediment.

3.2 Categories of Microorganism Fate in Stormwater

The fate of microorganisms (including both indicators and pathogens) in stormwater and sediment, as well as the interaction of microbes with suspended sediment and their subsequent deposition in the sediment, is considered in this review. The resuspension of sediment-associated microbes is also discussed.

Inactivation of microorganisms in sediment is observed when inactivated or stressed organisms are non-detectable by standard methods. This is the definition historically used in the literature to study inactivation rates. Scientists continue to debate what constitutes microbial death; however, this review does not address that issue. Predation, or biologically mediated removal from the water or sediment, occurs when predacious organisms (e.g. protozoa, zooplankton, or bacteria) consume other organisms. Sorption/desorption occurs when microorganisms in stormwater associate with or separate from particles, reversibly or irreversibly. Sedimentation takes place when stormwater microbes, either free or particle-associated, sink to the base of the water column where they are deposited under the influence of gravity. Resuspension occurs when microbes associate with sediment and move into the water column. Growth is an additional possible fate for bacterial microorganisms in stormwater.

3.3 Inactivation of Microorganisms in Stormwater

Microorganisms in stormwater have evolved to survive in a niche (the human intestine) that shares little in common with stormwater. Thus, the physio-chemical environment of stormwater is less than ideal for growth of microorganisms in stormwater (bacteria) or proliferation (virus and protozoa). The project team is unaware of any evidence that within the water column, microorganisms in stormwater can multiply.

The mechanisms whereby fecal indicators and pathogens become non-viable in the extra-enteric environment are not entirely clear. In general, the subject of bacterial "death" and senescence is the subject of intense research (Nystrom, 2004). Bacterial pathogens likely enter a prolonged stationary phase that may involve dwarfing, corruption of information flow from DNA to RNA to proteins, fragmentation of chromosomes, elevated mutation rates, and oxidative damage of proteins and membrane lipids. Viral pathogens may lose infectivity when their protein shell or nucleic acid suffers damage. Protozoan cysts and oocysts are more difficult to inactivate than bacteria and viruses because of their protective thick membrane, but damage to the macromolecules that compose them can cause inactivation.

In this section, the factors that influence the viability or infectivity of microorganisms in stormwater in the aqueous environment are described. The results of an inactivation experiment where bacterial microorganisms in stormwater are stressed, such as that conducted by Gauthier et al. (1989), may depend largely on the initial nutritional state of the organism. Although researchers seed most bacterial inactivation experiments with "starved" organisms, this may not always be the case. Therefore, caution is required when extending the results of lab experiments to the natural population in the environment. Further, the literature review did not identify any viability or infectivity studies that were conducted using stormwater as a medium. Thus, the focus is on studies conducted on microorganisms in fresh, estuarine and sea water, with and without physical and chemical augmentations. More research has examined fecal indicator organism viability as compared to pathogen viability and infectivity. This is due to a number of factors including the fact that fecal indicators play a central role in risk assessment for drinking water and bathing beaches. The factors explored in this section include sunlight, temperature, salinity, pH, and nutrient/metal concentrations.

3.3.1 Sunlight

For decades, researchers have documented the effect of sunlight on the viability of fecal indicator organisms, and to some extent pathogens. They have used laboratory and in-situ microcosm experiments, as well as field studies of polluted marine and freshwater beaches and estuaries. At Huntington State Beach, California, Boehm et al. (2002) observed a distinct diurnal signal in total coliform, *E. coli* and *Enterococcus* levels in the surf zone. During the dark hours of the night and early morning, fecal indicator bacteria are typically two to ten times higher than when the sun is out (Gauthier et al., 1989). At a freshwater beach on Lake Michigan, Whitman et al. (2004) found that *E. coli* levels are significantly lower on sunny versus cloudy days. The effect of sunlight is even evident in Antarctica near the sewage outfall at Rothera Point. During the austral summer when there are 24 hours of sunlight, Hughes (2003) determined that fecal coliform is present only within 150 m of the outfall, while during the austral winter, it extends over 500 m beyond the region sampled.

The exact mechanism whereby fecal indicator organisms and pathogens become non-viable after sunlight exposure is not entirely clear. Photons can excite exogenous or endogenous sensitizers present in the water and damage DNA or other cellular components directly. They can also cause damage indirectly by promoting the production of free radicals in the presence of dissolved oxygen and organics. Examples of exogenous sensitizers include humic acids. Carotenoids (fat-soluble pigments present in plants, algae, and photosynthetic bacteria) can reduce light-induced damage in bacteria, but these are typically absent in enteric organisms. Colloidal clay, which certainly is present in stormwater, may reduce light-induced damage by stabilizing excited sensitizers. Davies-Colley et al. (1999) demonstrated a relationship between the inactivation of *E. coli* and *Enterococcus* in sewage-spiked waste stabilization pond effluent and the amount of dissolved oxygen in the water, suggesting that the presence of free radicals promote inactivation of fecal indicator organisms.

When fecal indicator organisms undergo exposure to sunlight, they clearly lose culturability. It is less obvious whether they can regain their culturability via intracellular repair mechanisms. Oguma et al. (2001) showed that the formation of pyrimidine dimers during light-induced inactivation of *E. coli* accompany the organism's loss of culturability. When exposed to low levels of fluorescent light, *E. coli* repairs its damaged DNA and regains its colony-forming ability. Pommepuy et al. (1996) showed that enteropathogenic *E. coli* can enter a viable but non-culturable (VBNC) state when exposed to sunlight and lose the ability to be plate counted. However, they retain their ability to produce the toxin after their entry into the VBNC state. These findings suggest that sunlight-mediated inactivation of fecal indicator organisms in laboratory and field surveys may not be permanent.

Fujioka and Yoneyama (2002) showed that poliovirus (sabin type 1), coxsackievirus (type B5) and echovirus (type 7) (quantified by plaque assay) are more resistant to sunlight inactivation than *E. coli* and moderately more resistant than *Enterococcus*. Johnson et al. (1997) examined the survival of *Giardia*, *Cryptosporidium*, poliovirus and *Salmonella* in marine waters exposed to sunlight. *Giardia* cysts showed a 3-log reduction in three hours when exposed to sunlight, compared to 77 hours in the dark. *Cryptosporidium* oocysts proved to be hardier, with 10% surviving past 48 hours in sunlight compared to more than 97 hours in the dark. Poliovirus numbers were reduced approximately 1-log per 10 hours in sunlight compared to 1 –log per 20 hours in the dark. *Salmonella* was the least hardy of all organisms tested, with a 4-log reduction within 4 hours in sunlight. In the dark, 1-log reduction in *Salmonella* occurred after 24 h. Other researchers, including McCambridge and McMeekin (1981) and Evison (1988) examined *Salmonella* and compared its decay in sunlight to indicator organisms. Johnson et al. (1997) also showed that depressed water quality, as defined by turbidity and densities of *Enterococcus*, reduced the survivability of the organisms tested. Besnard et al. (2002) showed that exposure of Listeria monocytogenes to sunlight resulted in conversion to a VBNC state. Obiri-Danso and Jones (2001) showed artificial sunlight significantly and adversely impacted *Campylobacter* populations.

3.3.2 Temperature

In cooler waters, fecal indicator organisms can persist in a culturable state for longer periods than in warm waters. Solic and Krstulovic (1992) showed that the time required for a 90% reduction in fecal coliforms decreased 55% for each 10°C increase in water temperature. Field research as well as laboratory experiments supported these results. However, based on the coefficient of correlation and light and dark experiments, solar radiation is a more significant parameter than temperature. Noble et al. (2004) obtained similar results, finding higher inactivation at higher temperatures (20°C rather than 14°C) for total coliforms, *E. coli*,

Enterococcus and F+ specific coliphage. In addition, these researchers found that the overall inactivation rate was highest for total coliforms and lowest for *E. coli*. A study conducted by Serrano et al. (1998) in Spain confirmed that lower water temperatures were associated with higher concentrations of total coliforms, fecal coliforms, and coliphages. In another study, Halton and Nehlsen (1968) found that temperatures nearing 0°C increased the survival of *E. coli* in seawater. In experiments with coliphage, Berry and Noton (1976) found 5% of the initial phage numbers remained after 48 h of incubation at 4°C, whereas only 0.003% remained after 48 h at 14°C. Wait and Sobsey (2001) found similar results for *E. coli*. Burkhardt et al. (2000) showed that male-specific bacteriophages (MSB) correlated negatively with estuarine water temperatures. This study indicated that at cold temperatures (5°C), the MSB was a useful enteric virus indicator.

In a similar study conducted by Niemi (1975), the survival of the *E. coli* phage T7 in different water samples, averaged over winter and summer measurements, decreased to 10% in 16 days at 3°C, while it took almost four days at 20°C. Contrary to other studies mentioned, Nasser et al. (2003) found that *E. coli* seeded in seawater had a die-off of four orders of magnitude within six days regardless of the incubation temperature of 15°C and 30°C. Sinton et al. (1994) studied the inactivation of fecal indicator bacteria in the dark. They found that fecal coliforms from sewage effluent had a lower T_{90}[7] (82.3 hrs) at warm temperatures (15-20°C) than at colder temperatures (8-10°C) with a T_{90} of 115 hours.

Similarly, effluent from a meat processing plant had a T_{90} of 59.1 and 121 hours for warm and cold temperatures, respectively. *Enterococcus* followed a similar trend for sewage effluent, with a decrease of T_{90} from 461 to 288 hours for cold and warm temperatures, respectively. Comparing cold to warm water experiments, the meat processing plant effluent also had a decrease in T_{90} *Enterococcus* from 288 to 256 hours. Vasconcelos and Swartz (1976) also found that *E. coli*, streptococcus and *Salmonella* persistence decreases with increasing temperature. They noted that Streptococcus faecalis was more persistent than other bacteria in this study.

A limited number of researchers have explored pathogenic bacteria inactivation as a function of temperature. Their results are somewhat consistent with those reported for fecal indicator organisms. Wait and Sobsey (2001) showed that *Shigella* sonnei and *Salmonella* typhi survive longer in low water temperatures without pressure from other stressors. *Campylobacter* spp. also survive for extended periods (up to four months) at low temperatures (Rollins and Colwell, 1986; Thomas et al., 1999; Obiri-Danso and Jones, 2001). Evison (1988) did a comparative study on the survival of various fecal indicators (*E. coli*, fecal streptococci, coliphages MS 2 and f 2) and bacteria pathogens (*Salmonella* typhimurium 12, 12a, 110, *Salmonella* anatum, *Shigella* sonnei, *Shigella* sonnei, Yersinia enterocolitica, and *Campylobacter* fetus) at temperatures of 2°, 5°, 10°, 15°, 20°, and 25°C. Most organisms showed worse survivability at high temperatures as compared to low temperatures. Changes in temperature affected *Campylobacter* the least. Y. enterocolitica showed increased growth rates at high temperatures. *Salmonella* anatum showed better survivability at 5°C, 10°C, and 15°C than at 5°C, 20°C, and 25°C. Reddy et al. (1981) found that the die-off rate for some bacterial indicators

---

[7] Time required for a 90% inactivation in number of organisms.

and pathogens approximately doubled with a rise in temperature of 10°C between 5-30°C. These organisms included fecal coliforms, *E. coli*, Aerobacter aerogenes, Streptococcus faecalis, *Salmonella* typhimurium, Enterobacter aerogenes, and poliovirus.

Wait and Sobsey (2001) showed that some viruses (poliovirus type 1 and parvovirus) have 90% lower inactivation times at 28°C than at 6°C, following a pattern similar to bacteria in which the inactivation rates are lower at lower temperatures. Similarly, Chung and Sobsey (1993) found greater survival rates for hepatitis A virus, poliovirus, and rotavirus at 5°C, compared to 25°C. Lo et al. (1976) showed that poliovirus 1, echovirus 6, and coxsackievirus B5 survived for an extended period at low temperatures compared to high. At 4°C, viruses remained infectious for up to 46 weeks. Nasser et al. (2003) found that coxsackievirus A9 showed dependence on temperature for inactivation, with the concentration decreasing two orders of magnitude at 15°C after 30 days, compared to a decrease of five orders of magnitude at 30°C in only 15 days. O'Brien and Newman (1977) saw a pattern of increasing inactivation with increasing temperature. Poliovirus 1, poliovirus, and coxsackievirus showed a 1-log reduction in 19, 3, and 7 h, respectively, at temperatures between 23°C and 27°C. In contrast, at temperatures ranging from 4°C to 8°C, the 1-log reduction took place in 46 hours for poliovirus 1 and 58 hours for coxsackievirus. Fujioka et al. (1980) illustrated another example of increased virus inactivation at higher temperatures. Poliovirus type 1 showed a 3-log reduction in two days at 24°C, while the concentration at 4°C remained stable for three days.

Protozoan parasites are more resistant to temperature changes. In Nasser et al. (2003), *Cryptosporidium* seeded into stream water and incubated for 30 days at 15°C and 30°C were not inactivated at all. Fayer et al. (1998) found that at a constant salinity of 10 ppt, *Cryptosporidium* retained their infectivity after 12 weeks at temperatures of 10°C and 20°C. Other studies, such as that by Robertson et al., (1992), have shown that *Cryptosporidium* can survive very cold temperatures. In this study, *Cryptosporidium* was subject to slow freezing. After 21 hours at a temperature of -21°C about 67% of oocysts became inactive, and after 152 hours, *Cryptosporidium* had reached 90% inactivation.

3.3.3 Salinity

Salinity may affect pathogen inactivation in coastal receiving waters such as estuaries or the coastal ocean. Solic and Krstulovic (1992) found an inverse relationship between fecal coliform survival and salinity concentrations, where T90 decreased as salinity increased. Specifically, the researchers found an approximate 55% reduction in T90 for each 5% salinity increment in the 7-15‰ salinity range, and a decrease of about 15-40% in T90 for each 15% salinity increment in the 15-40‰ salinity range. They also noted that higher salinity and high levels of solar radiation combined to produce a synergistic effect, resulting in higher mortality rates of fecal coliforms.

A study conducted by Hanes and Fragala (1967) showed that *E. coli*, coliforms and *Enterococcus* had greater inactivation rates with increasing salinities, with 6.2, 4.6, and 1.6 times greater death rate, respectively, at 100% seawater compared to 0% seawater. Similarly, Anderson et al. (1979) found a decreased survival rate for *E. coli* with increasing salinity, ranging from 53.5% survival at 10 ppt after eight days of exposure, to 2% survival at 30 ppt for the same period of time. Fujioka et al. (1981) studied the comparative stability of fecal coliform and fecal streptococci. The researchers found that seawater caused rapid inactivation of fecal streptococci and fecal coliform (T90 of 36 and 98 min, respectively), whereas the organisms remained stable for three days in freshwater.

In a comparative study of pathogens, Evison (1988) found that salinity measurements in the dark had different effects on different organisms at 15°C. *Shigella* flexneri, *Salmonella* typhi and *Yersinia enterocolitica* showed growth or prolonged survival (negative T_{90}) in freshwater, as compared to more saline waters. In contrast, fecal streptococci had the highest survival rate in 3.5 ppt salinity water. In addition, the author mentions that of the salinities studied (0, 0.44, 0.88, 1.31, 1.75, 2.63, and 3.5 ppt[8]), a salinity of 0.88% generally gave the best survival rates. *Yersinia enterocolitica* actually grew at low salinities.

Johnson et al. (1997) found that at oceanic salinities (35 ppt), *Giardia muris* inactivated more rapidly than it did at lower salinities of 28 ppt and 8 ppt. In addition, the researchers found that salinity and light accomplished 99% inactivation of cysts within 3-6 hours, as opposed to 121.3 and 97.1 hours in canal and marine water, respectively, in the dark. This result agrees with inactivation results found for fecal indicator organisms. The same study found that *Cryptosporidium* survived longer in saline waters. Fayer et al. (1998) found a similar trend, reporting that inactivation rates for *Cryptosporidium* decreased with increased salinity. Fayer found that at 20°C, *Cryptosporidium* remain infectious for 12 weeks incubated in water with salinities of 0 and 10 ppt, and two and four weeks in salinities of 20 and 30 ppt, respectively.

Enriquez et al. (1995) compared the survivability of adenovirus, hepatitis A, and poliovirus 1 in seawater and freshwater, finding that adenovirus outlasted the other two viruses. In another virus study, Akin et al. (1976) found the rate of loss in infectivity of poliovirus 1 in waters varied as a function of salinity. Poliovirus 1 showed a reduction of approximately 3, 4, and 5 logs for salinity levels of 1, 20 and 10 g/kg respectively. The authors noted that their results did not agree with a similar study where the loss of infectivity of poliovirus was directly proportional to salinity.

3.3.4 pH

Limited research has been performed on the effect of pH on survivability of microorganisms in stormwater. Solic and Krstulovic (1992) noted that fecal coliforms thrived at a pH range of 6 and 7. Outside of this range, fecal coliforms declined rapidly, with a slightly greater mortality in acidic environments. The only pathogen inactivation study that examined the effect of pH is that of LaBelle et al. (1980), who showed that virus stability seemed to increase within the pH range of 7.8 and 8.4 and decrease outside this range.

3.3.5 Nutrients

Few studies have systematically examined the effect of nutrients (including dissolved organic carbon and trace metals) on the persistence of microorganisms in stormwater in the environment. Thomas et al. (1999) found *Campylobacter* did not survive longer in nutrient-

---

[8] Evison (1988) reported that the salinity ranges investigated were 0 to 3.5 ppt. However, it may be possible that there is a typo in the article and the reported salinities are in parts per hundred.

containing microcosms. In microcosm studies in fresh and salt waters, Noble et al. (2004) found that nutrient levels had an insignificant effect on the persistence of fecal indicator bacteria or fecal indicator viruses. Evison (1988) measured varying responses for different concentrations of nutrients (0.025, 0.25, 2.5 and 25%) in the form of sterile raw sewage at a temperature of 15°C. At relatively high concentration of nutrients (25%), Evison (1998) observed growth or prolonged survival of *Shigella* sonnei and *Campylobacter* fetus in both fresh and sea water. In contrast, Evison reported that *Shigella* typhimurium 12a and *Salmonella* anatum showed higher survival rates, but no growth under these conditions.

The stress level or phase of seed organisms may affect the results of inactivation studies. Gauthier et al. (1989) showed that *E. coli* in seawater previously grown in nutrient broth for different periods of time had variable inactivation rates. After incubating *E. coli* in nutrient broth for three days, the researchers found that the concentration of *E. coli* decreased from approximately 105.5 to about 10-1 CFU/mL in five days. After 25 days in nutrient broth, *E. coli* decreased in concentration from 106.5 to about 106 over 5 days.

3.4 Predation of Microorganisms in Stormwater

Predation puts pressure on both microorganisms in stormwater and indigenous microbes. Protozoa, nanoflagellates, microflagellates, ciliates, amoeba, and bacteria may prey upon viruses as well as other protozoa and bacteria. Barcina et al. (1997) asserted that "grazing by protozoa is one of the main biological processes that control allochthonous bacterial density." Fujioka et al. (1980) documented anti-viral microorganisms in natural waters. Literature documenting the importance of lytic phage in the destruction of stormwater bacteria in water, could not be located, hence the discussion of this process is omitted.

Enzinger and Cooper (1976) showed that estuarine waters from San Francisco Bay contained both protozoa (including ciliates and amoeba) and lytic bacteria (including Bdellovibrio bacteriovorus) that cause declines in *Enterococcus*. In these waters, protozoa exerted greater grazing pressure on *Enterococcus* than did the lytic bacteria. In a similar study, McCambridge and McMeekin (1980) showed that protozoa and lytic bacteria reduced *Enterococcus* levels from 108 to 10 CFU/mL over 10 days, a much greater reduction than previously thought possible via this mechanism. Anderson et al. (1983) further confirmed the importance of eukaryotic predators in the decline of *Enterococcus* by using diffusion chambers in an estuary. Roper and Marshall (1978) found that the addition of clay particles, often present in stormwater, to a suspension containing *Enterococcus* and protozoa significantly reduced predator-*Enterococcus* interactions. They explained this observation as the clay particles acting to "shield" bacteria from predation. McCambridge and McMeekin (1981) confirmed the importance of predation on other fecal bacteria suspended in estuarine water, including Klebsiella pneumoniae, *Salmonella* typhimurium, Streptococcus faecium, and Enterobacter aerogenes. The presence of sunlight exacerbates predation pressure on *Enterococcus* McCambridge and McMeekin, 1981; Rhodes and Kator, 1990) even though sunlight has no effect on the numbers of predators present in the waters. Chamberlain and Mitchell (1978) suggest that solar radiation causes injury to *Enterococcus* that makes them susceptible to predation. Rhodes and Kator (1988) reported first-order decay constants between 0.43–0.92/day and 0.48-0.87/day for *Enterococcus* and *Salmonella* spp., respectively, in estuarine water. Rhodes and Kator attribute 90% of the decay to the presence of organisms unfiltered by a 0.2 μm pore size filter. Menon et al. (2003) confirmed that grazing of three fecal bacteria *E. coli*, Streptococcus faecium, and *Salmonella* typhimurium was responsible for 90% of their overall mortality rates in fresh river water and coastal marine waters. There may be some preference amongst grazers for particular enteric organisms. For example, Macek et al. (1997) found that ciliates from Mexican waters preferred Vibrio cholerae over *Shigella* spp. and *Salmonella* typhi. However, Vibrio cholerae is an autochthonous organism, unlike *Salmonella* typhi.

Stott et al. (2003) showed that free-living protozoa could consume *Cryptosporidium* oocysts at a rate of 1.9 oocysts per protozoa per hour. Stott et al. (2001) reported that protozoa also ingested other protozoan parasites such as microsporidia. The fate of the ingested pathogen is not clear; they may re-enter the environment via excretion in fecal pellets.

Gonzalez and Suttle (1993) determined that protozoan or zooplankton grazers also consume viruses and virus-like particles, although clearance rates in their experiments for viruses were 4% of those measured for bacteria. Gonzalez and Suttle also showed that small viruses, approximately the size of picornaviruses, rotaviruses and adenoviruses, were ingested more slowly than larger viruses (those on the order of 100 nm). Fujioka et al. (1980) hypothesized the existence of yet unidentified virus-inactivating microorganisms in seawater. Researchers suggested that these organisms might secrete material that inactivates enterovirus, coxsackievirus B-4, and echovirus 7. Although Fujioka et al. (1980) could not identify similar organisms from a limited number of regional freshwater streams, such organisms may exert an influence on enteric viruses in stormwater, especially if the receiving water is marine.

3.5 Sorption/Desorption of Microorganisms in Stormwater and Suspended Sediment

This section describes the mechanisms of sorption and the consequence of particle-associated microorganisms in stormwater on fate of microorganisms in stormwater. Paerl (1975) hypothesized that sorption of microbes to particle surfaces benefits the microorganisms, providing them with higher concentrations of nutrients as well as protection from predation, phage, and inactivation. Thus, particle-microbe interactions may have a major impact on the fate of microorganisms in stormwater in the environment.

3.5.1 Mechanisms of Sorption

The physio-chemical and biological properties of organism surfaces and the size and chemical characteristics of particulate surfaces govern interactions between microorganisms in stormwater and particles. Steric, electrostatic, and hydrophobic interactions may contribute to interactions between particles and microorganisms in stormwater. Steric interactions arise when the macromolecules that compose the organism's surface extend into a solution where they may interact with particulate surfaces. Electrostatic interactions between particles and a microorganism in stormwater occur when the surfaces of the organism takes on a significant negative or positive charge. The charge and its magnitude are controlled strongly by solution and surface chemical conditions. Ionic strength, pH, and metal concentrations influence electrostatic interactions. Maier et al. (2000) found that most particulate surfaces have a net negative charge in natural waters, as do most viruses and bacteria. However, heterogeneity on particle surfaces can result in patotal coliformshes or edges that are positively charged. The classic example is kaolinite, which Rand and Melton (1977) found to have positively charged edges. Thus, one must examine electrostatic interactions and van der Waals forces for specific solution and sediment chemistries. Hydrophobic interactions describe the tendencies of non-polar, organic groups to interact in aqueous media. These interactions are expected to be important when both microorganisms in stormwater and surface contain numerous non-polar moieties. They are considered to be among the primary forces controlling bacteria-particle interactions.

Significantly, Walker (2005) noted that bacteria-particle interactions are more difficult to predict than virus or (oo)cysts since the hydrophobicity of a cell surface changes as a function of nutritional state. The presence of bacterial pili, fimbriae, or flagellae may also control bacteria-particle interactions. Belas and Colwell (1982) established this relationship for Vibrio cholerae. Palmateer et al. (1993) found that some bacteria also excrete mucopolysaccharite, a chemical that has the effect of cementing bacteria to the sorption surface. Researchers have determined that lipopolysaccharides (LPS) and extracellular polymeric substances (EPS) play an important role in bacterial adhesion to surfaces (Kannenberg et al., 2001; Abu-Lail and Camesano, 2003; Walker, 2005).

It is possible that microorganisms in stormwater that are adsorbed to particles can be desorbed. Marshall (1975) defined adsorption as "reversible" if weak van der Waals forces control sorption. However, if shear forces are strong enough, even "irreversibly" bound microorganisms in stormwater may be unbound from particles. Thus, the association between microorganisms in stormwater and particle should not be viewed as permanent, especially in stormwater where solution chemistry is variable and fluid velocities and shear rates are dynamic.

3.5.2 Observations of Associations between Microorganisms in Stormwater and Suspended Particles

Only a handful of studies have directly examined the interaction between microorganisms in stormwater and particles in suspension in a field setting. All but one of these observed only indicator organisms. The most common method for direct examination of particle-microbe interaction is through physically separating size classes (via centrifuging, sieving, or filtering) and subsequently assaying those size classes for organisms.

Characklis et al. (2005) specifically examined the percent of indicator organisms associated with suspended particles that were "settleable" (including fecal coliforms, *E. coli*, *Enterococcus*, *Clostridium perfringens*, and coliphage). They found that between 30% and 55% of the non-spore-forming bacteria, from 50-70% of C. *perfringens*, and between 20% and 60% of coliphage were associated with particles during storm events. The values were slightly lower during background non-storm conditions. However, one must interpret these results cautiously, as the authors did not attempt to desorb bacteria from suspended sediments once they separated the size fractions. Other researchers found that 20% of fecal coliform (Schillinger and Gannon, 1985) and between 5% and 100% of fecal coliform and *Enterococcus* (Mahler et al., 2000) associated with particles in stormwater. Borst and Selvakumar (2003) found that concentrations of microorganisms in stormwater all increased in water samples after physical agitation, a process that dislodges microbes from particle surfaces.

Research on particle-associated suspended microbes has been conducted in media other than stormwater. Pommepuy et al. (1992) showed that in an estuary in Spain, between approximately 5% and 75% of culturable fecal coliform were associated with particles greater than 3 μm in diameter. These authors did not attempt to desorb bacteria from suspended sediment; they separated the free microbes from the particle-associated fraction using filtration. The more turbid the waters, the more bacteria were sediment associated. Rao et al. (1993) showed that bacterial concentrations vary in differently sized suspended particulates with maximum populations in the 20-40 μm-sized fraction. Shibata et al. (2004) found that *Clostridium perfringens* concentrations at two beaches in Florida correlated significantly to turbidity, although they did not specifically check that the microbes were particle-associated. Mallin et al. (2000) also observed a relationship between turbidity and fecal coliform in coastal creeks.

Palmateer et al. (1993) made the only observations of pathogens sorbed to suspended particles. They found up to 500 *Salmonella* per mm2 of suspended sediment in agricultural drains, utilizing microscopy. They used a direct method of enumeration and did not require culture-based assays.

Although there are limited field studies on helminths, Schwartzbrod et al. (1989) found that helminth eggs have a tendency to be particle-associated.

Researchers have developed empirical relationships between the concentration of microbes sorbed to suspended sediments and the total number of microbes in suspension for modeling purposes. Jamieson et al. (2005) modeled the fate of free and particle-associated *E. coli* in an alluvial stream and tested the model against field observations. They were able to model the sorption of free microbes to particles as irreversible, using a partitioning fraction F between 0.20 and 0.44. They defined:

$$F = \frac{SS \times C_{sed}}{C_{total}}$$

where SS is the suspended solids in a sample (g), C_{sed} is the concentration of bacteria on sediment (CFU/g) and Ctotal is the total number of bacteria in the sample. Bai and Lung (2005) modeled sorption of fecal bacteria to suspended particles using a linear isotherm and assumed a portioning coefficient between 10-4 and 10-6 L/mg.

These studies demonstrate the importance of the association of microbes with suspended sediment. The variability in the size classes of particles with which organisms are associated and the percent of the total associated microorganisms in stormwater is predictable. Site- and water-specific characteristics determine sorption and desorption. Therefore, it is unlikely that two locations would have the exact same amount of stormwater microbes associated with suspended material.

3.6 Sedimentation/Deposition of Microorganisms in Stormwater

Either attached to particles or free in aqueous solution, microorganisms in stormwater may settle to the base of the water column and be incorporated into the sediments (Jamieson et al., 2002; Steets and Holden, 2003). This process is also termed deposition. Numerous studies have reported high concentrations of human pathogens and indicator organisms in sediments of polluted waters, and have investigated the organisms' fate within the sediments, over the last 40 years (Gerba and McLeod, 1976; Goyal et al., 1977; Van Donsel et al., 1967; Smith et al., 2001). Researchers have also observed an increase in numbers of microorganisms in stormwater (including fecal coliforms, *Salmonella*, coliphage, *Clostridium perfringens*, *Enterococcus*, Aeromonas, *Giardia*, *Cryptosporidium*, and enterovirus) in sediments after rain or storms (Craig et al., 2004; Ferguson et al., 1996; Lipp et al., 2001a; Smith et al., 2001). However, documenting the process of pathogen and indicator deposition in the sediments has proven challenging. This is because the fractal nature of aggregates makes it difficult to characterize the settling velocity of free or particle-bound microbes in stormwater as a function of particle size, shape, and density. Currents opposed to the direction of gravity, as may be induced by turbulent eddies, prevent sedimentation from occurring. They are also difficult to model and characterize.

Researchers have made direct observations of deposition in the field and in laboratory flume settings. Rittenberg et al. (1958) observed rapid sedimentation of particle-associated coliform bacteria near a sewage outfall, which effectively resulted in self-purification of the water column. Obiri-Danso and Jones (2000) conducted deposition studies of *Campylobacter*, fecal coliforms, and fecal streptococci in situ in coastal England. They found that during tidal cover, between 0 and 1.3 MPN per square sediment of these microbes were deposited. Searcy et al. (2005) conducted laboratory-scale flume experiments showing deposition of *Cryptosporidium* oocysts in streambeds. Their results show that hydrodynamic interactions between the stream and water in the stream bed, including hyporheic exchange, force free and particle-associated *Cryptosporidium* oocysts to accumulate in the sediments. They also show that advective deposition of oocysts is just as important, or more important, than gravitational sedimentation of oocysts to the bed. Settling velocities of the oocysts (both free and particle-associated) averaged between 0.8 and 53.3 μm/s, depending on the type of particles present during the experiment. Searcy et al. (2005) showed that despite its negative charge, *Cryptosporidium* interacted strongly with kaolinite and illite clays, iron oxide colloids, and natural sediments. Further, these researchers showed that the protozoa settled with the colloids it had adhered to, resulting in deposition. The mechanism of adhesion is believed to be van der Waals forces. Their results were not significantly affected by changes in solution pH and ionic strength. In contrast, Dai and Boll (2003) found that *Cryptosporidium* and *Giardia* did not interact with soil particles. However, these researchers used very low concentrations of soil in their experiments.

Some researchers have examined deposition by using settling velocities and shear stress at the bed as fitting parameters for models and actual field observations of dynamics of microorganisms in stormwater found in lakes or streams. Bai and Lung (2005) used a single settling velocity of 0.0001 m/s and a critical shear stress for deposition of 0.2 N/m2 to fit observations of coliform in a stream to their model. Jamieson et al. (2005) used settling velocities between 0.00003 and 0.0003 m/s to model *E. coli* sedimentation in a stream. In a mass balance, Auer and Niehaus (1993) deduced a settling rate of 1.38 m/d for particle-associated fecal coliform in a lake.

3.7 Fate of Microorganisms in Sediments

Researchers have acknowledged since at least the 1960s that sediments can be reservoirs of pathogens and indicator organisms. While some researchers view sediments to be a "dead end" in the life cycle of enteric microbes (Winfield and Groisman, 2003), others have found that some have extra-enteric lifecycles (Byappanahalli et al., 2006). The recovery of pathogenic bacteria (including, but not limited to, Vibrio vulnificus (Lipp et al., 2001a), *Salmonella* (Hendricks,1971; Goyal et al., 1977), and *Campylobacter* (Obiri-Danso and Jones, 1999), protozoa (Lipp et al., 2001b), and viruses (Lipp et al., 2001a) from the sediments when they could not be recovered from the water column, or at a higher concentration than in the water column, suggests that sediments should be viewed as an important reservoir for microorganisms in stormwater.

Several studies have specifically identified the effect of storms and rain on the numbers of microorganisms present in sediments. These studies have observed an increase in the numbers of microorganisms in stormwater (including fecal coliforms, *Salmonella*, coliphage, *Clostridium perfringens*, *Enterococcus*, Aeromonas, *Giardia*, *Cryptosporidium*, and enterovirus) in sediments after rain or storms (Craig et al., 2004; Ferguson et al., 1996; Lipp et al., 2001; Smith et al., 2001). After a large flood in Germany, Strauch et al. (2005) recovered pathogens and indicator microbes from sediments for up to one year.

Researchers have investigated the reason microorganisms in stormwater can be found in high concentrations in sediments compared to the water column, primarily through microcosm studies. The literature suggests that inactivation and predation are reduced in the sediments, while growth is possible. Sediment particles shield the microorganisms from adverse biological, physical, and chemical conditions that are present in a chemically variable aqueous phase. They also provide increased nutrients that bacteria may use to reproduce (Gerba and McLeod, 1976). Some of these studies and their findings are outlined below. Little if any work has examined the persistence of protozoa in sediment microcosms and therefore the fate of these organisms in sediment will not be discussed.

3.7.1 Bacteria

Studies have shown that microorganisms in stormwater adsorbed to both organic and inorganic sorbents in sediments extend their survival periods from a few days to several months. These particles act as physical shields for sunlight, desiccation, antibiotics, and predation (Marshall, 1975; Davies and Bavor, 2000; Desmarais et al., 2002; Jin et al., 2004). In microcosm studies, Anderson et al. (2005) showed that the inactivation rates of indicator bacteria are slower in the sediments compared to the water column. Another benefit of sorption is the buffering capacity of some sorbents like clay, which provides a more suitable environment for acid-sensitive bacteria (Stotzky, 1966; Stotzky and Rem, 1966). Sediments potentially harbor osmolytes that can aid in the regulation of microbial osmotic pressure (Gauthier et al., 1991). Additionally, regrowth of enteric bacteria in some instances is possible (LaLiberte and Grimes, 1982; Marino and Gannon, 1991; Hardina and Fujioka, 1991). This is because the nutrients that are readily available for bacterial uptake can be 10-100 times higher on the particulate's surfaces than in the surrounding water column (Paerl, 1975). Various studies have attributed the survival of fecal coliforms in sediments to the concurrent presence of soluble organics (Hendricks, 1971; Enzinger and Cooper, 1976; Gerba and McLeod, 1976; Verstraete and Voets, 1976; Tate, 1978; Novitsky, 1983; Marino and Gannon, 1991; Davies et al., 1995). Anaerobic conditions in sediments also reduce potential predation for certain microbes (Enzinger and Cooper, 1976).

Gerba and McLeod (1976) showed that sediments contain a large amount of leachable nutrients that promote the growth of *E. coli*. They showed that removing the effects of predation permits the maintenance of even higher densities of *E. coli*. Craig et al. (2004) found that *E. coli* persisted for a longer period when it was placed in sediments with relatively small particle sizes and high levels of organic carbon. However, Pommepuy et al. (1992) questioned the hypothesis that microbes persist in sediments because of better availability of nutrients. They point out that much of the organic matter in sediments may be refractory and difficult for bacteria to utilize.

The moisture content of sediments and soils plays a major role in extending the persistence of *E. coli* (Byappanahalli et al., 2006). A study of *E. coli* O157:H7 showed that soil moisture, commonly associated with soils with fine particles, correlated positively to the persistence of the organism (Mubiru et al., 2000). Howell et al. (1996) also showed that fine sediments prolong the persistence of fecal coliforms and fecal streptococci.

Maki and Hicks (2002) showed that various suspended particles (clay, silt, and organic particles) do not affect the survivability of *Salmonella typhimurium* in freshwater. Although they did observe organism attachment, they reported that it did not affect the viability of the cells.

3.7.2 Viruses

LaBelle and Gerba (1980) found that sediment provides protective physical sites for virus adsorption. Chung and Sobsey (1993) found that sediment also protects viruses from adverse survival conditions such as unfavorable temperature. Many researchers have studied virus-particle interactions extensively in connection with virus transport in the subsurface, and have observed that in aqueous environments, most viruses are particle associated. Gantzer et al. (1994) showed that in the presence of high levels of montmorillonite in high ionic strength seawater, poliovirus 1 adsorbed strongly to particles and experienced increased survivability. LaBelle and Gerba (1979) demonstrated that several enteroviruses and a rotavirus bound strongly to estuarine sediment. They determined that the addition of organics such as humic acids or wastewater effluent did not influence adsorption. Only echovirus appeared to be reversibly adsorbed to sediments. Rao et al. (1984) found that poliovirus, coxsackie B virus, and echovirus preferentially attached to suspended solids rather than being suspended in the water phase or associated with the nephloid layer and compacted sediments in Galveston Bay, TX. The researchers also found that poliovirus and rotavirus adsorbed to sediments retained their infectivity longer (19 days) than viruses freely suspended in water (9 days). Le Guyander et al. (1994) found hepatitis A and enterovirus in sediments of a naturally contaminated coastal shellfish harvesting area. Some evidence suggests that viral genomes may remain intact longer than infectious virus in the presence of particles. This possibility could make results from gene-based detection assays for human virus difficult to interpret in samples of particle-laden stormwater.

Many studies document the presence of human virus in sediments of contaminated areas (e.g. Kueh and Grohmann 1989). Chung and Sobsey (1993) showed that poliovirus, rotavirus, and hepatitis A virus (HAV) had T99.99 of 1899, 234 and infinite days in sediment at 5°C, compared to 37, 871, and 84 days in seawater at 5°C. The single-stranded RNA viruses survived longer in sediments than in the water column, while the rotavirus (double-stranded RNA virus) did not.

3.8 Resuspension of Microorganisms in Stormwater

Pitt and Field (1990) found that if a microorganism settles to the base of the water column, it can be incorporated into the sediments where it may remain indefinitely, or until the sediment is scoured (i.e., resuspended) by shear force at the bed. Scouring of microorganism-laden sediments may represent a long-term source of microorganisms to waterbodies. Indeed, some work has found that water quality violations only occur when a sediment reservoir of fecal organisms is present (Crabill et al., 1999).

Sediment composition and cohesiveness determine the critical stress that causes resuspension of bottom sediments. Bai and Lung (2005) set a critical shear stress at the bed to 0.4 – 0.5 N m^{-2} to model scouring and resuspension of microorganisms in a creek and a lake. They found that the shear stress critically affected whether microbes were transported from the sediments to the water column.

A variety of physical processes, including wave action, currents, and human activity, can affect the shear stress at the bed. Sleath (1984) observed that in the surf zone of beaches, breaking waves cause most of the sediment resuspension. At a beach in New Zealand, researchers found that passing waves resuspended *Enterococcus* in the sediments. Solo-Gabriele et al. (2000) found that high tides wash *E. coli* from creek sediments to the water column in subtropical Florida. They based their conclusions on intensive field observations that showed the covariation of tide height with *E. coli* in the water. Plummer et al. (1987) showed that tidal currents in an estuary periodically resuspended particle-associated bacteria. Whitman and Nevers (2003) found a significant correlation between *E. coli* in the water column and the behavior of wind and waves, known to cause resuspension of lake sediments. Crabill et al. (1999) documented increased fecal coliforms in an Arizona creek during increased flow regimes and increased visitor activity when sediments were high in fecal coliforms. Grimes (1980) showed that human activities in the form of dredging caused large influxes of fecal coliforms from the sediment to the water column. Jamieson et al. (2004) seeded the sediment of alluvial streams with *E. coli* and found that "diffusion" of *E. coli* from the sediments to the water column occurred during both base-flow and storm conditions. During base flows, only *Enterococcus* concentrations increased and not suspended solid concentrations downstream of the seeded sediment. This suggests that changes occurred in bacteria adsorption to sediments and interstitial flow rates (rather than in scouring of sediments). During storms, both *E. coli* and suspended sediment concentrations increased, suggesting resuspension of both sediment and microbes. Jamieson et al. (2004) also found that when stream sediments contained more clay, bacteria were less likely to be released and transported downstream.

It is believed that the resuspension of sediments in surface waterbodies that is caused by increased stream flows during storm events often accompanies a rapid deterioration of water quality. Consequently, the occurrence of scouring suggests that the concentration of microorganisms in stormwater may not necessarily reflect the true degree of microbial contamination of stormwater runoff. Levels of microorganisms in stormwater in stormwater could be on the threshold of being a health risk concern upon entering rivers, while scouring in microbial-rich sediments may significantly elevate the microbial concentration of the stormwater runoff. This was observed by various researchers (Hardina and Fujioka, 1991; Crabill et al., 1999) who found water quality only deteriorated when sediments rich with *Enterococcus*, *E. coli*, or fecal coliforms were resuspended.

Consequently, depending on the degree of resuspension along the runoff pathway, the concentration of microorganisms in stormwater runoff may not be consistent. Therefore, the location of sampling sites used to assess concentrations of microorganisms in stormwater, relative to areas of potential sediment-associated microbes, is one of the key factors that should be noted and taken into account during interpretation.

3.9 Summary and Conclusions

Microorganisms in stormwater may enter a waterbody in the free phase or adsorbed to a suspended particle. Once in the waterbody, microbes may be preyed upon, become inactivated, adsorb/desorb from particles, or be deposited within sediments at the bottom of the water column. Microorganisms in sediments are exposed to a different environment than they are in the water column. In sediments, predation, inactivation and growth rates are more favorable to organism survival than in the water column. When sediments are subjected to high shear stress due to waves, currents, or tides they may be scoured and thus act as a source of microorganisms in stormwater to the water column.

Inactivation results from various mechanisms such as lysis or mutation of genetic material. It is mediated by warm temperatures, sunlight, low nutrients, or high salinity. Non-spore-forming bacteria and viruses are more readily inactivated by these stressors than protozoan pathogens and spore forming bacteria. The influence of nutrients and metals on an organism's inactivation/persistence is not straightforward, probably because the nutritional requirements of many stormwater bacteria are not completely understood. On the other hand, the effects of temperature and salinity have been examined by numerous researchers. Additionally, an increase in the association between microorganisms in stormwater and particles, either suspended or deposited, is believed to decreases inactivation rates. In some cases, the growth of bacteria has been observed in sediments. Evidence suggests that particles provide nutrients and shielding from harsh environmental conditions to stormwater microbes. The experimental methods used to determine inactivation rates have not been standardized, so extension of inactivation rates from studied to unstudied systems should be done cautiously. Of special concern are the findings of Gauthier et al. (1989) showing that the nutritional state or phase of a seed pathogen may influence its inactivation rate.

Predation of microorganisms in stormwater has proven to be extremely important in reducing their numbers in waterbodies. Bacteria, protozoa, and viruses have all been shown to be removed, to varying extents in natural waters. The association of microorganisms in stormwater with particles or their deposition in the sediment has been shown to reduce predation pressure.

Observations of up to 100% of fecal bacteria or coliphage bound to suspended sediment in waterbodies have been reported in the literature. Modelers have taken various approaches to explaining this observation, including assuming adsorption is an equilibrium process and occurs following classical sorption isotherms. The deposition of microbes to sediments can occur whether or not microorganisms are sediment associated. The process of deposition and its role in microorganism fate deserves more study because it is an important link between water column and sediment. It is extremely important to understand the conditions under which microorganisms are deposited in the sediment because they can survive for extended periods and may even be able to reproduce there. Sediment represents an environment where microbes are sheltered from predation and environmental conditions conducive to inactivation. Numerous studies have documented the increased persistence of microbes there. In addition, studies have shown that it is possible to isolate pathogens from sediments when they cannot be isolated from the water column. Thus sediment, particularly after a storm, represents a potentially important environmental reservoir of pathogens. Sediments, and associated microorganisms, can be resuspended when a critical shear stress is experienced at the bed. The required stress to cause sediment or microbe resuspension will depend on properties of the sediment including its size and cohesiveness. In general, a critical shear stress may be exceeded during the passage of waves (including tidal waves) and during increased stream flow.

3.10 Recommendations

Future work on fate of microorganisms in stormwater should focus on the movement of microorganisms in stormwater between the fluid- and particle-associated phases. In particular, work to further understand particle-microbe interactions in the water column, deposition and resuspension are needed.

Stormwater monitoring programs should include sediments as a sampling matrix. If monitoring is to be done when concentrations of microorganisms in stormwater are highest, then sampling should occur at night when there is little sunlight and the water is cold. Sampling should be conducted when bottom sediments would potentially be scoured and resuspended in the water column. When assaying for microorganisms in stormwater, care should be taken to account for microorganisms in stormwater that may be particle-associated.

CHAPTER 4.0

SOURCE TRACKING METHODS FOR MICROORGANISMS IN STORMWATER

1. Which waterborne pathogens pose the greatest risk to human health and what are the concentrations of these pathogens and indicator organisms observed in stormwater?
2. Are there defensible relations between indicator organisms and pathogens of public health concern in stormwater?
3. What is the fate of stormwater microorganisms (i.e., pathogens and indicator organisms) in the environment?
4. What microbial source tracking (MST) techniques are used to identify sources and assess potential public health risk?
5. What are the most effective stormwater microorganism control technologies and what are the estimated costs?
6. What methods have been historically used to quantitatively characterize potential human health risk associated with exposure to waterborne pathogens and what is the most appropriate method for stormwater?
7. What critical data are needed to enable the assessment of risk associated with exposure to pathogens in stormwater and what is the recommended approach?

4.1 Introduction

Given that regulations were written using currently validated methods for enumerating indicator microorganisms of fecal contamination, there is a need to further refine these measurements to distinguish those organisms that may have originated from animals from those of humans. Applications for this refined indicator measure include the characterization of sources of fecal contamination in watersheds and in recreational waters. It has been this desire that has led to the development of the science of microbial source tracking (MST).

This chapter provides an overview of the rationale behind MST, descriptions and limitations of the various MST methods, and discusses the outstanding issues and future research needed to resolve these issues. Although they are not discussed in this report, it is important to note that other methods based on analyzing physical and chemical constituents have been developed to identify sources or categories of sources of microorganisms in stormwater (Pitt et al., 1993; Brown et al., 2004).

4.2 Rationale and Applications of Microbial Source Tracking

Microbial source tracking (MST) studies are carried out in watersheds that affect drinking water sources and other beneficial uses. The purpose is to better define the source of microbes of fecal origin. What is the "source" in microbial source tracking? In a watershed, this can be wildlife, domestic animals, and human activity (e.g., leaking septic systems).

Why conduct analysis for these sources? If a source of fecal contamination can be linked to domestic animals (e.g., cows, dogs), watershed management practices may be instituted to reducing the impacts of these types of animals on water quality. If the source can be determined to be from wild animals (e.g., squirrels, geese, ducks), again there may be management practices instituted to reduce their direct impact on the receiving waters. If human sources can be identified, then management practices may be put into place to control these inputs (e.g., repairing or re-positioning septic systems).

Therefore, it is thought that by knowing the source of fecal contamination in a watershed that feeds a drinking water source or a recreational waterbody, these inputs can be controlled and limited. For drinking water, the drinking water treatment plant would therefore have less potential disease-causing organisms to remove from the source water, thus reducing the risk of infectious disease through the drinking water route. For recreational waterbodies, the risk of infectious disease could be reduced by implementing source controls or best management practices (BMPs).

4.3 Microbial Source Tracking Methods

Ideally, MST is based upon the belief that a direct correlation exists between an organism found in the environment and the specific animal source of that specific organism. However, MST has also been used in a broader context to define a group of microorganisms that may have originated with groups or sub-groups of host organisms. A literature review was undertaken to summarize the state of the science for microbial source tracking. The literature search resulted in over 75 abstracts of published papers, which were reviewed for relevancy. From that group, 34 papers were retrieved and summarized as part of this chapter.

Many extensive reviews have been written on this topic (Bernstein et al., 2002; Scott et al., 2002; Simpson et al., 2002; Griffith et al., 2003; Myoda et al., 2003; Atherholt, 2004; Stoeckel et al., 2004). In addition, there have been several industry and government sponsored workshops and method comparison studies devoted to MST (Griffith et al., 2003; Stoeckel et al., 2004; WERF, 2005).

Microbial source tracking can be classified into three primary areas of method application: 1) host specific; 2) phenotypic; 3) genotypic. Host specific methods attempt to isolate microorganisms that are only found in specific hosts (e.g., humans). In contrast, phenotypic methods rely on the demonstration of specific metabolic traits in isolated organisms, while genotypic methods analyze the DNA of an isolate.

These primary areas of method application can be further broken down by whether they are “library-independent” (LI) or, “library-dependent” (LD) in nature. Library-independent methods are based upon detection of the target organism that is directly linked to the specific source. For example, if certain species of Bacteroides are found only in humans, then the detection of these types of bacteria would indicate a human source of fecal contamination. Likewise, if a certain species of Rhodococcus is only found in certain animals, and not humans, then the detection of this type of bacteria would indicate an animal source of fecal contamination (Long et al., 2002). Library-dependent methods are based upon characterizing microorganisms that are shared by animals and humans, but can be differentiated by characterizing either a phenotypic trait specific to that host-specific strain, or, by subtle genotypic differences. For example, *E. coli* are found in most warm-blooded animals; however, it is thought that human strains or subtypes can be differentiated from animal types by comparing specific nucleic acid sequences within their respective DNA. A summary table of the target organisms and methods used in MST is presented below in Table 4-1.

Table 4-1. Organisms and Methods Used in Microbial Source Tracking.

Host Specific Organisms (LI)	Phenotypic Methods (LD)	Genotypic Methods (LD)
Bacteroides spp. phage *Bacteroides* spp. Bacteroides-Prevotella group *Bifidobacteria* spp. *Clostridium perfringens* Coliphage *Escherichia coli* *Enterococcus* spp. Fusobacterium Peptococcus Rhodococcus Ruminococcus *Streptococcus* spp. Human or Animal-specific viruses	Antibiotic Resistance Analysis (ARA) Carbohydrate Utilization Profiling (CUP) *Escherichia coli* serotyping	Amplified fragment length polymorphisms Coliphage genotyping Denaturing gradient gel electrophoresis Enterotoxin biomarkers Multilocus sequence typing Plasmid typing Pulsed field gel electrophoresis Random amplification of polymorphic DNA Repetitive fragment length polymorphism (BOX, Rep, & ERIC- PCR) Ribotyping Terminal restriction fragment length polymorphism Sequencing organism-specific genes

Scott et al. (2002) and WERF (2005)

4.3.1 Library-Independent Microbial Source Tracking Methods

As discussed above, Library-Independent (LI) methods focus on the detection of those organisms that are thought to be specific to a specific host. Bernhard and Field (2000), utilize those bacteria of the Bacteroides-Prevotella group (also known as the Order Bacteroidales) to differentiate human fecal contamination from that of ruminants. This method was used in a watershed study to distinguish the contribution of fecal pollution of grazing cows from possible human inputs. Since these organisms are anaerobic and very difficult to culture from the environment, this molecular method is PCR-based and targets molecular markers of the Bacteroidales. Likewise, Bonjoch et al. (2004) targeted specific molecular markers of the Bifidobacterium, believed to be the most common bacteria type found in the human intestine. Ram et al. (2004) reported using PCR to detect the gene marker for MUG[9]-reactive organisms. This study revealed that there was geographic diversity with MUG positive *E. coli* present in two relatively close areas. Therefore, if one were to use this method for tracking *E. coli*, there may be too much variability to apply this method to more than one watershed.

Coliphage have been categorized such that certain types have been associated with human sources and animal sources. Vinje et al. (2004) were able to differentiate "F+" (male-specific) coliphage of humans from those of animals based upon the subtle genetic differences. By targeting these differences using a PCR-based method, the presence of these phage could be detected directly from environmental samples. Noble et al. (2004) used human enteric viruses and indicator organisms together to differentiate between human and non-human sources of fecal contamination. One of the limitations of using specific human pathogens as a target is that the

[9] MUG is a substrate that only certain microorganisms can utilize, specifically, *E. coli*

population under investigation may or may not be shedding or even harboring the targeted pathogen.

4.3.2 Library-Dependent Microbial Source Tracking Methods

Library-dependent (LD) MST methods can be further characterized as "phenotypic" or "genotypic". Phenotypic methods rely on measuring or detecting an expressed trait of the target organism. On the other hand, genotypic methods rely on detecting a specific nucleic acid sequence(s) for the specific target organism. Individual isolates must be recovered from various sources, e.g., fecal samples from animals in a watershed and stool samples (or sewage) from humans in the immediate area. The individual isolates are measured for the target characteristic. Data collected by either method are stored in a database and the measured characteristic(s) (phenotypic or genotypic) is compared to those from other isolated organisms from the various sources. The discussion that follows describes the different methods.

4.3.2.1 Phenotypic Methods

Typically, phenotypic methods are based on measuring the presence of certain enzymatic/metabolic activity. The individual isolates are exposed to a variety of substrates and the outcome of how the organism reacts to the substrate is documented. For example, a series of nutritional substrates are introduced individually to the isolate and if the organism utilizes the substrate, there is a measurable reaction. A profile of the types of substrates used characterizes that specific isolate.

Gamo and Shoji (1999) used the BIOLOG (Hayward, CA) identification system with multiple carbon sources to profile isolates from environmental communities. However, Smalla et al. (1999) concluded that phenotypic methods employing nutritional substrates may not accurately describe a microbial community. Only those organisms that are physiologically capable of utilizing a substrate(s) will be determined while other organisms of interest that are present, but not physiologically capable, will go undetected.

Ebdon et al. (2004), Hagedorn et al. (1999) and Harwood et al. (2003) described methods of categorizing the antibiotic resistance patterns of individual isolates. The individual isolates were applied to agar plates and antibiotic disks were added; the resistance (or susceptibility) of the isolates is assessed and the pattern is recorded. The type of antibiotics used is important; some antibiotics are used almost exclusively in animals (domestic and agriculture) and others are used almost exclusively in humans. Therefore, if an isolate has more resistance to antibiotics used in humans, the isolate is classified as being from a human. Isolates that match resistance patterns with antibiotics used with domesticated animals are thought to be from that source. Isolates that do not match either group be may classified as from "wildlife," or a population of animals that have not been exposed to antibiotics. This method relies heavily not only on the type of antibiotic used, but the dose of antibiotic in the assay. Earlier antibiotic profiles used about 12 antibiotics, but more recent profiles use 20 or more with three concentrations per antibiotic.

4.3.2.2 Genotypic Methods

Genotypic methods are similar to phenotypic methods in that a library of information is required for comparisons of isolate. As can be seen in Table 4-1, there are many ways to ascertain unique genetic sequences of microbial DNA. Again, these methods rely on the isolation of individual organisms, however, instead of demonstrating a specific metabolic trait, the DNA of the isolate is analyzed. Table 4-2 lists the advantages and disadvantages of these methods.

Table 4-2. Advantages and Disadvantages of Some Genotypic Methods.
Reprinted with permssion: Scott, T. M., Rose, J.B., Jenkins, T.M., Farrah, S.R., Lukasik, J. Microbial source tracking: current methodology and future directions. Appl. Environ. Microbiol. 2002, 68, 5796-5803.

Method	Advantages	Disadvantages
Pulse Field Gel Electrophoresis (PFGE)	Extremely sensitive to minute genetic differences	May be too sensitive to broadly discriminate for MST
PCR (BOX, Rep, ERIC)	Rapid, easy to perform	Reproducibility; reference database required; geographically specific
Ribotyping	Highly reproducible; classify isolates from multiple sources	Labor-intensive; reference library required; geographically specific

Pulse field gel electrophoresis has been in use for many years in public health investigations of foodborne outbreaks (Johnson et al., 1995). This technique produces highly resolved "fingerprints" of specific sequences of target DNA in order to discern subtle differences between compared sequences from different organisms. The U.S. FDA has used PFGE to match those isolates from contaminated foods to those recovered from ill patients. This technique has been applied to environmental isolates as well (Refsum et al., 2002; Steele et al., 1998).

PCR techniques have been employed in the detection of rarely occurring genetic sequences for almost twenty years. In this instance, it is used to amplify specific sequences directly from environmental samples, usually without isolating specific microorganisms. These sequences are then resolved to detect target organisms and/or compare the differences within the same species of organisms. Johnson et al. (2004) and Seurnick et al. (2002) have described applications of PCR-based methods where specific areas within a host DNA is amplified and compared to other isolates. The "repetitive-PCR" (Rep, BOX and ERIC) methods rely on targeting areas in the DNA that are repeated in several locations, however, there may be subtle differences within species that can be detected (Stern et al., 1984; Hulton et al., 1991; Versalovic et al., 1991).

Ribotyping relies on separation and analysis of specific well-conserved DNA fragments to identify a specific strain. Ribotyping requires isolated organisms from the study area for developing a library of organisms from that area. Parveen et al. (1999) was the first to apply ribotyping to distinguishing isolates from various sources. Hartel et al. (2002) and Wheeler et al. (2002) employed ribotyping of isolates to investigate the strain differences between isolates from different geographic areas.

4.4 Limitations of MST Methods

No method is perfect, and that is certainly the case with the currently employed MST methods. The limitations of MST methods are discussed in the following sections.

4.4.1 Library-Independent Methods

Library-independent methods that are dependent upon the direct isolation or detection of specific microorganisms (e.g., coliform or *Enterococcus* assays) have been the traditional methods of determining microbial water quality. The difficulty with isolation methods, such as culturing, is that they rely on the presence of physiologically active organisms that can grow in the in-vitro culture environment. In many cases, the microbes of interest are physiologically stressed once they leave their animal host and, while present, are non-culturable. Some of the culture conditions may not be practical or are too expensive for use in regular monitoring. For example, the cultivation of anaerobic bacteria (bacteroides, bifidobacteria, clostridia, etotal coliforms) requires strict anaerobic conditions of an anaerobic culture chamber. Animal viruses require specific animal tissue culture hosts. Only a small number of tissue culture assay systems exist relative to variety of viral pathogens. In addition, culture methods for almost all pathogens are ill defined and/or not validated. Finally, pathogens occur in relatively small number, are physiologically stressed and difficult to capture, separate from background organisms, and cultivate from environmental samples.

PCR methods that are designed to directly detect the presence of specific microorganisms have been developed and applied. The limitations of this technique include:

- presence of naturally occurring PCR-inhibiting agents in the sample;
- possible cross-reaction with heretofore unknown organisms;
- choice of specific nucleic acid sequences ("primers");
- optimization of PCR assay conditions;
- stringency of the confirmation assay (DNA probe or DNA sequencing procedures);
- inability to differentiate between living and dead organisms; and,
- stringent laboratory quality control measures to reduce the occurrence of false positive outcomes.

4.4.2 Library-Dependent Methods

4.4.2.1 Phenotypic Methods

Phenotypic methods are based on measuring the presence of certain specific enzymatic/metabolic traits. Measuring a specific expressed trait in environmental bacteria is a primary issue since the specific traits may not be expressed by the organism. While the organism may have the necessary metabolic components in place, it may not be able to operate in the manner desired by the investigator, due to environmental conditions. Furthermore, Smalla et al. (1999) reported that this technique did not accurately measure the metabolic activity of the numerically dominant species in a community.

In the case of antibiotic resistance assays the isolates may be specific to a geographic area and any library has to be kept up-to-date throughout the year to track temporal variations. In addition, there is no way to classify those organisms that do not provide a profile (e.g., susceptible to all antibiotics). In general, methods that use molecular markers (e.g., Bacteroides-Prevotella bacteria) are of limited value because little is know about their distribution in the environment and survival characteristics. In addition, it is not known if these organisms are host specific (Scott et al., 2002).

4.4.2.2 Genotypic Methods

The genotypic methods rely on the construction of large libraries of isolates from the various sources. Thus, the genotypic approach relies on sampling known sources such that bacteria later recovered from a water sample (the "unknown") can be classified based on comparison to the known source. Some controversy exists regarding the size of genetic library needed to resolve differences between isolates (see Data Analysis, below). Hartel et al. (2002) has recognized that large (> 2000 isolates) libraries are needed for the application of ribotyping isolates[10].

Most genotypic methods are based upon the clonal nature of bacteria. As bacteria divide and multiply, the same genetic information is passed from mother cells to daughter cells. However, the stability of the bacterial genome varies from microbe to microbe. For example, many of the genotypic libraries that have been created are based upon Escherichia coli. Barnes and Gordon (2004) have contended that *E. coli* has too much variability, even within clonal lines. Alternatively, researchers have suggested that bacteria in the *Enterococcus* group may present a more stable line (WERF, 2005). McLellan et al. (2003) found high genetic diversity in *E. coli* isolates and suggested that *E. coli* characterization may be most feasible within a limited geographic area, e.g., for watershed specific studies. Application of this approach to the identification of host sources of *E. coli* in the environment requires more extensive information on the overall genetic structure of the natural populations, particularly the host specificity of strains, as well as net temporal changes in the population and geographical differences in strain occurrence (McLellan et al., 2003).

Seurnick et al. (2002) used a rep-PCR method to conduct MST studies and they concluded that to make this approach widely applicable, the possible temporal diversity, geographical limitations, and survival of these source-specific genotypic markers in the environment need to be assessed. Likewise, Johnson et al. (2004) stated that their results further indicated that other important issues must also be resolved to understand more fully the potential applications and limitations of this and other library-based microbial source-tracking methodologies. Among these are questions concerning the inclusion of identical DNA fingerprints from the same animal in the library and the number of fingerprints that must be included in an *E. coli* known-source library to adequately capture the diversity of *E. coli* genotypes that exist among potential host animals. Ultimately, one must determine whether *E. coli* exhibits a sufficient level of host specificity to allow unambiguous assignment of unknown environmental *E. coli* isolates to specific host animals (Johnson et al., 2004).

4.4.2.3 Data Analysis Limitations

For library-dependant methods, it is necessary to apply statistical analysis to be able to match and contrast the data added to the library. The limitations of the statistical analysis must be understood in order to interpret the results more accurately. There have been a few review articles on the appropriate analysis that can be carried out on these types of data (Albert et al.,

---

[10] Hartel, at a WERF workshop on MST (February, 2005), stated that he has given up on genomic typing techniques due to the overall uncertainties with and high costs of conducting these types of studies.

2003; Ritter et al., 2003; Barnes and Gordon, 2004). These researchers warned against overly simplistic commercially available statistical software packages that will have their own inherent limitations, such as not allowing for the manual manipulation of statistical parameters.

Albert et al. (2003) suggested that the creation of a useful library of fingerprints should be a dynamic process where new fingerprints and groups are periodically included and the library is re-analyzed. Upon analysis, it may be found that the specificity of the fingerprints is not high enough to separate source groups. New features must be evaluated (geographic/temporal separation, other organisms or type of data) to improve specificity.

Ritter et al. (2003) carried out a comprehensive review of the statistical methods used in MST. They found no consensus as to which methods were the most appropriate. They assessed discriminant analysis, nearest neighbor analysis, maximum similarity and average similarity, along with several measures of distance or similarity. The researchers also examined threshold criteria for excluding uncertain or poorly matched isolates from final analysis for their ability to reduce false positives and increase prediction success. They studied six independent libraries constructed from indicator bacteria isolated from fecal materials of humans, seagulls, cows and dogs. Three of these libraries were constructed using the rep-PCR technique and three relied on antibiotic resistance analysis (ARA). Five of the libraries were constructed using *E. coli* and one using *Enterococcus* spp. Overall, the outcome of this study suggests a high degree of variability across statistical methods. Despite large differences in correct classification rates among the statistical methods, no single statistical approach emerged as superior. Thresholds failed to consistently increase rates of correct classification and improvement was often associated with substantial effective sample size reduction. Simply changing the type of statistical analysis alone could re-classify an isolate from human to animal.

4.5 Outstanding Issues

Atherholt (2004) has concluded that all MST methods require further refinements. The goals and scope of MST studies should be carefully defined. Future studies might improve pollution source assignation by employing both phenotype and genotype tests simultaneously. However, methods require standardization. Chemical marker methods used in conjunction with MST may provide a more clear assessment.

As discussed above, *E. coli* may not be best suited for library-based methods. Barnes and Gordon (2004) used mathematical models to show that there is enough strain variation in each host such that detection in environmental samples may not actually reflect the source host. There are also variations within each host that may overlap with other sources (hosts). Barnes and Gordon (2004) argued that, regardless of the coliform species being monitored, the success of any effort to trace the source of fecal contamination depends on several assumptions being valid. One of these assumptions is that the clonal composition of the species isolated from a waterbody represents the clonal composition of the species in the host populations responsible for the fecal inputs into the environment.

In method comparison studies, the results indicate that the methods rarely agreed with each other and, in many cases, the results could not be replicated with the same isolates. Stewart et al. (2003), Griffith et al. (2003), Harwood et al. (2003), Myoda et al. (2003), Noble et al. (2003) and Stoeckel et al (2004) all reviewed a method comparison study that was sponsored by the Southern California Coastal Water Research Project. Some of the conclusions from those studies were:

- Most methods correctly identified samples containing human fecal contamination.
- Samples without human fecal contamination were more difficult to assess, with high false positive rates for the presence of human fecal contamination.
- None of the methods was able to identify the dominant source in all samples.
- Every method performed poorly in identifying all sources of contamination in the sample.
- However, the comparison studies were not designed for optimum performance for all of the methods tested.

Another area that has not been fully documented is the cost of conducting a MST study. Typically, library-based methods can range from $20,000-200,000 depending upon the size of the watershed and the variety of sources within the watershed (Linda O'Connell, CA State Water Resources Control Board, personal communication).

4.6 Summary and Future Research

The project team undertook a literature review to summarize the state of the science for microbial source tracking (MST). MST depends on the assumption that a direct correlation exists between an organism found in the environment and the animal source of that specific organism. There are also chemical measures of animal and human activity that have been used in source tracking. However, this review was limited to those techniques that utilize microorganisms.

There are a number of MST techniques that have been proposed, all with their specific advantages and limitations. In summary, there is a confused history of the application of the currently employed techniques. Certainly, there is no one way of performing a MST study, as it appears as though the nature of the targeted watershed will dictate the application of the various techniques. Suggested improvements for MST include (Stewart et al., 2003):

1. Decide upon which bacterial species will provide best discrimination between sources;
2. Improved selection of enzymes and antibiotics for library based methods;
3. Definition of the library size needed for the specific method chosen;
4. Application of appropriate algorithms for the chosen method (library-based).
5. The development of quantitative library-based methods;
6. Selection of the appropriate combination of methods;
7. Standardization and validation of MST methods;
8. Development of quality assurance program and quality control measures; including the use of analytical standard materials; and,
9. Development of an open-access database with the posting of libraries from MST studies.

It is believed that the future of MST lies in the suggested improvement #6 (above). One should not rely solely upon microbial measures, but should try to combine with chemical source tracking and general watershed assessment methods.

CHAPTER 5.0

TREATMENT CONTROLS FOR MICROORGANISMS IN STORMWATER

1. Which waterborne pathogens pose the greatest risk to human health and what are the concentrations of these pathogens and indicator organisms observed in stormwater?
2. Are there defensible relations between indicator organisms and pathogens of public health concern in stormwater?
3. What is the fate of stormwater microorganisms (i.e., pathogens and indicator organisms) in the environment?
4. What microbial source tracking (MST) techniques are used to identify sources and assess potential public health risk?
5. What are the most effective stormwater microorganism control technologies and what are the estimated costs?
6. What methods have been historically used to quantitatively characterize potential human health risk associated with exposure to waterborne pathogens and what is the most appropriate method for stormwater?
7. What critical data are needed to enable the assessment of risk associated with exposure to pathogens in stormwater and what is the recommended approach?

5.1 Introduction

Various studies have assessed the ability of structural treatment controls (e.g., detention basins, retention ponds, bioretention, infiltration basins, stormwater wetlands) to reduce pollutant concentrations and loadings in stormwater. However, research on the efficiency of these treatment controls, or Best Management Practices (BMPs), has historically concentrated on physical and chemical contaminants such as total suspended solids, nutrients and metals. Studies that have documented the removal of microorganisms in stormwater (i.e., microbial indicators and pathogens) by BMPs are limited and use inconsistent methodologies that make comparisons difficult. Additionally, available performance information is generally related to fecal indicators (total and fecal coliforms, *E. coli*, and *Enterococcus*), not to specific pathogens (Smith and Perdek, 2004). Despite these challenges, available information on the efficiency of structural treatment controls to reduce microorganisms in stormwater was compiled and the results are summarized in this chapter.

The purpose of this review is to assess the known utility of structural treatment controls to reduce pathogen and/or indicator organisms in stormwater. This was accomplished through a review of available literature on technical performance, specific control projects and research endeavors. The review addresses issues with BMP performance data and acknowledges climatic, site, and other variables that may affect pollutant removal efficiency. Additionally, construction and on-going maintenance and operation costs estimates are provided. Information was collected primarily from published sources, although gray literature was also reviewed where appropriate. The BMP categories listed in Table 5-1 were included in this review.

Table 5-1. Best Management Practice (BMP) Categories Included in the Literature Review.

BMP Categories	Alternative Name	Description
Detention Basins	♦ Dry pond ♦ Extended detention basin ♦ Detention pond ♦ Extended detention pond	Dry basin designed to completely empty at some time after stormwater runoff ends. Adaptation of the detention basins used for flood control. The term "dry" implies that there is no significant permanent water pool between storm runoff events.
Retention Ponds	♦ Retention pond ♦ Stormwater ponds ♦ Retention pond ♦ Retention basin ♦ Retention extended detention pond	Ponds that have a permanent pool of water, unlike detention basins, which dry out between storms. The permanent pool of water is replaced in part or in total by stormwater during a storm event. The design is such that any available surcharge capture volume is released over time.
Infiltration Basins & Trenches	♦ Infiltration galley	Basins that can capture a given stormwater runoff volume and infiltrate it into the ground, transferring this volume from surface flow to groundwater flow. Infiltration trenches can be generally described as a ditch filled with porous media designed to encourage rapid percolation of runoff to the groundwater.
Vegetated Swales	♦ Grassed channel ♦ Dry swale ♦ Biofilter ♦ Retention swale	A vegetated swale is a shallow grass-lined channel with zero, or little, bottom width designed for shallow flow near the source of storm runoff.
Stormwater Wetlands	♦ Constructed wetlands ♦ Shallow marsh ♦ Wetland system ♦ Pocket wetland ♦ Gravel based wetland ♦ Wetland basin	Similar to a retention pond (with a permanent pool of water) and more than 50% of its surface covered by emergent wetland vegetation. Depth is typically shallow (< 1 meter).
Sand Filters	♦ Austin sand filter ♦ Delaware sand filter ♦ Washington D.C. sand filter	BMP that uses sand or a sand mixture, with or without a presettling basin, to remove a fraction of the constituents found in stormwater.
Proprietary Devices	♦ Gravity Separation ♦ Swirl Concentration ♦ Filtration	Includes proprietary devices Stormceptor®, BaySaver®, Vortechs™, Downstream Defender™, and various other catch-basin inserts, filtration devices, and hydrodynamic separators.
Disinfection	♦ Chlorination/Dechlorination ♦ Ultraviolet Light Irradiation ♦ Ozonation	Chemical or physical processes that destroy pathogenic organisms. Common disinfectants include chlorine, ultraviolet light, chlorine dioxide and ozone.

Sections 5.2 through 5.8 provide brief summaries of the pollutant removal efficiencies, application and costs of implementing BMPs listed in Table 5-1. Section 5.9 summarizes the information presented in the chapter.

5.1.1 BMP Effectiveness Terminology

The term BMP effectiveness is used loosely in the literature reviewed by the project team. Definitions vary drastically between studies, making comparisons difficult. In order to better clarify the terminology used to describe the level of treatment achieved by a BMP type, terms used throughout this chapter are defined in Table 5-2. These definitions are consistent with those described in the International Stormwater BMP Database.

Table 5-2 Definition of Terms Used to Evaluate the Level of Treatment Achieved by a BMP.
(Reprinted with permission from ASCE. Strecker, E., Quigley, M., Urbanas, B.R., Jones, J.E., Clary, J.K. Determining urban storm water BMP effectiveness. J. Wat. Res. Plan. Manag. 2001, 144-149)

Term	Definition
Best Management Practice (BMP)	A device, practice, or method for removing, reducing, retarding, or preventing targeted stormwater runoff constituents, pollutants, and contaminants from reaching receiving waters.
BMP System	A BMP system includes the BMP and any related bypass or overflow. For example, the efficiency can be determined for a offline retention basin either by itself (as a BMP) or for the BMP system (BMP including bypass).
Performance	A measure of how well a BMP meets its goals for stormwater that the BMP is designed to treat.
Effectiveness	A measure of how well a BMP system meets its goals in relation to all stormwater flows
Efficiency	A measure of how well a BMP or BMP system removes pollutants.

The initial focus of this literature review was to determine effectiveness of BMPs and BMP systems in removing microorganisms in stormwater. Effectiveness was chosen as the key measurement because it accounts for stormwater flows that are treated and those that bypass the system, due to engineering constraints. However, much of the literature does not accurately report bypass volumes, and therefore treatment control efficiency (as opposed to effectiveness), which only considers the stormwater actually being treated by the structural control, became focus of the literature review. Summaries contained within this report are therefore optimistic with regard to assessing the level of treatment achieved by a BMP.

5.1.2 Challenges in Comparing BMP Efficiency Studies

A variety of factors make comparisons of BMP studies difficult and caution should be used when comparing the BMP studies presented in this report. A few of these factors are briefly described below.

5.1.2.1 Standards and Protocols

Although some projects conducted to date have done an excellent job of describing the effectiveness of specific BMPs and BMP systems, there is a lack of standards and protocols for conducting BMP assessment and monitoring work. Inconsistent study methods, lack of associated design information and reporting protocols make wide-scale assessments difficult, if not impossible (Strecker et al., 2001). For example, individual studies often include the analysis of different constituents and utilize different methods for data collection and analysis, as well as report varying degrees of information on BMP design and flow characteristics. Most studies reviewed do not rely on statistically designed sampling and data analysis protocols. The differences in monitoring strategies and data evaluation alone contribute significantly to the range of BMP efficiencies that have been reported in literature to date.

5.1.2.2 Storm Variability

Representative monitoring of stormwater BMP performance is not a simple and straightforward task. The principal challenge facing persons implementing BMP monitoring programs is the temporal and spatial variability of stormwater flows and pollutant concentrations (U.S. EPA , 2002b). Stormwater quality at a given location can significantly vary both between storms and during a single storm event (U.S. EPA, 1983; Soller et al., 2004a) making it difficult to characterize BMP efficiency. Some studies have used event mean concentrations (EMCs) to better represent the average concentrations in influent and effluent. However, samples are still typically collected using a grab sample methodology and may not be representative of the range of concentrations observed during a storm event. Therefore, it is important to note that a monitoring project which collected a small number of samples is not likely to provide a reliable indication of the efficiency of a given BMP and could produce erroneous conclusions.

5.1.2.3 Site and BMP Characteristics

Efficiency data may vary based on event and seasonal characteristics, design and siting disparities and variations in watersheds. Frequently, it is difficult to compare studies due to differences in factors such as drainage area size, degree of watershed imperviousness, the volume and rate of runoff entering the structural control, and the surrounding land use of the contributing drainage area. Additionally, the designs of structural controls are often dissimilar, such as the ratio of the structural control, volume to the contributing drainage areas, the retention time in the BMP, and the physical dimensions and the construction of the control. This further confounds comparisons of data.

5.1.2.4 Monitoring Logistics

A variety of frequently utilized structural controls (e.g., infiltration trenches and infiltration basins) are rarely monitored to evaluate efficiency. The primary reason behind the relatively sparse amount of data available on the efficiency of these and other structural controls is the complexity inherent in gathering samples of inflow and outflow to determine pollutant removal. With regard to infiltration BMPs, water routinely discharges over an extensive surface area into adjacent soil layers, making collection of a representative outfall sample difficult.

5.1.2.5 Loads Versus Concentrations

Pollutant removal efficiency estimates for BMPs can vary depending on the type of data reported. BMP data are typically reported in one of two ways, concentration and/or loading. As previously mentioned, methods used to determine the concentrations of pollutants in BMP influent and effluent may vary from a single grab sample to EMCs. Loading estimates are typically developed by multiplying concentration by volume. Relative to microorganisms in stormwater, the data are typically reported as concentrations to depict the range of efficiencies for BMPs. However, loads are generally believed to be a more meaningful indicator in BMP efficiency in nearly all cases.

5.2 Detention Basins

Detention basins, also known as dry ponds, dry extended detention ponds, detention ponds, and extended detention basins are basically depressions in the ground that temporarily store a portion of stormwater runoff following a storm event to allow particles and associated pollutants to physically settle in the basin (Barrett, 1999; CASQA, 2003). Detention basins differ from retention ponds in that they lack a permanent water pool between storm events.

Historically, detention basins have been successful in reducing peak discharges of stormwater to receiving waters to limit downstream flooding and channel erosion. However, not until the 1960s were detention basins designed for stormwater treatment (U.S. EPA, 2002b). Basins began to be redesigned to extend the detention time beyond that required for flood control to provide increased water quality benefit. Thus, the term "extended" detention basin was coined.

Since that time, research has demonstrated that detention basins are viable and effective treatment facilities for physical and chemical contaminants (Barrett, 1999; WMI, 1997; Emmerling-Dovino, 1995). When properly designed, substantial reductions are possible in the total suspended sediment load and constituents associated with these sediments (U.S. EPA, 2004). However, substantial uncertainty remains in the efficiency of these BMPs to reduce microorganisms in stormwater.

5.2.1 Removal Efficiencies

The following section provides a summary of available data on the efficiency of detention basins to remove microorganisms in stormwater. Direct comparison of these data were not possible since watershed area, drainage time, number of storms monitored, pond design, and sampling techniques differed for each study. However, the information presented in these sections provides an example of the range of detention basin efficiencies. The studies are presented in chronological order, beginning with the most recent.

The California Department of Transportation (CalTrans) conducted a BMP retrofit study in Southern California in which five detention basins were installed (CalTrans, 2004). CalTrans results from sampling a total of about 55 storm events are shown in Table 5-3. In summary, CalTrans reported that influent and effluent concentrations were not significantly different ($p>0.05$).

Table 5-3. Detention Basin Removal Efficiencies for Microorganisms in Stormwater.

Study	Parameter	Mean EMC (unless otherwise noted)		% Removal	Log Removal	Significance (p < 0.05)
		Influent (MPN/100ml)	Effluent (MPN/100ml)			
CalTrans (2004)	Fecal Coliforms	900	2000	-122	-0.35	0.607
USGS (2004)	Fecal Coliforms	45	53	-15	-0.07	NR
Harper et al. (1999)	Fecal Coliforms	3412	115	97	1.47	NR

NR = Not Reported

A small project was conducted in the U.S. Virgin Islands in 2003 to evaluate the efficiency of detention basins as BMPs for fecal coliform reduction under varying climatic conditions (USGS, 2004). A total of eight water quality samples were collected at the inlet and outlet of the detention basin during three storm events. The authors indicated that BMP efficiency was low, based on percent removal of fecal coliforms (Table 5-3). However, it is important to note that mean EMCs for fecal coliforms at the inlet were relatively low (45 MPN/100 ml), compared to other studies summarized in this chapter.

Investigations were also conducted by Harper et al. (1999) from 1997 to 1998 to evaluate the pollutant removal efficiency of a detention basin system in DeBary, Florida. A total of 35 separate storm event samples were collected and analyzed over the 6-month sampling period for fecal coliforms. The authors reported a 98% reduction in fecal coliforms during the study (Table 5-3). The authors suggested that the high efficiency of the BMP was attributable to a filter underdrain system which assisted in the removal of bacteria attached to total suspended solids (TSS).

5.2.2 Application in Urbanized Areas and Costs[11]

Detention basins have a broad range of applications in the urban landscape, although it is imperative to evaluate (prior to installation) if a detention basin is applicable at a specific site. In most cases, detention basins should be employed at sites having drainage areas of at least four hectares to allow for an adequate volume of water entering the basin during storm events (CWP, 2001). Thus, the utilization of detention basins in highly urbanized areas tends to be less practicable, due to land availability.

Construction costs for detention basins are generally less expensive, on a cost per unit area treated basis, than retention basins, sand filters or stormwater wetlands. Generally, the chief capital cost for installing a detention basin is the cost of excavating the site (Barrett, 1999). Additionally, with regard to all structural controls, the key factor affecting the cost of installing the control is whether it will be applied as in conjunction with a new construction or is a retrofit development. General construction cost estimates ranges between \$11 and \$34/m^3 (CWP, 2001; CalTrans, 2004).

Additionally, as with all structural BMPs, routine maintenance and inspection activities are also required in addition to integrating elements into the detention basin design to reduce the frequency and scope of maintenance needed. Wayne County (2001) estimates annual maintenance costs to be roughly \$10,000 for each detention basin maintained. Maintenance costs include such activities as inspections, debris and sediment removal, erosion control, and structural repairs. Annual maintenance costs estimated by CalTrans (2004) were \$3,132/basin, which included inspections, maintenance, vector control, administrative duties, and materials. Alternatively, the CWP (2001) estimates that the annual maintenance costs of a detention basin are roughly three to five percent of the construction cost.

5.3 Retention Ponds

Retention ponds, also known as wet ponds, stormwater ponds, wet retention ponds, retention basins and retention extended detention ponds, are constructed basins that have a permanent pool of water, unlike detention basins, which dry out between storms. The permanent pool of water is replaced in part or in total by stormwater during a storm event. Pollutant removal

---

[11] Costs estimates for all BMPs are reported as found in the literature reviewed, and are not normalized for geographical location or adjusted for any given year.

in retention ponds can occur through a number of mechanisms. The removal of suspended solids and associated constituents largely occurs through gravity settling (i.e., sedimentation), but the presence of aquatic plants and microorganisms can also provide uptake of nutrients and degradation of organic contaminants.

5.3.1 Removal Efficiencies

The following section briefly summarizes available data on the efficiency of retention ponds to remove microorganisms in stormwater. Direct comparison of these data were not possible since watershed area, sampling methodology, number of storms monitored, pond design, and removal efficiency techniques differed for each study. However, the information presented from these studies does provide a general range of pollutant reduction that can be expected from retention ponds. The studies are presented in chronological order, beginning with the most recent.

CalTrans (2004) assessed the pollutant removal efficiency of one retention pond in Southern California that collected highway runoff. Influent and effluent grab samples were collected during storms and analyzed for fecal coliforms. Mean fecal coliform EMCs reported were 11,700 MPN/100mL and 100MPN/100mL for the influent and effluent, respectively, equating to a 99 % (2.0 log) removal (Table 5-4). However, due to the small number of storms sampled and high variability in concentrations, influent and effluent results were not statistically different ($p = 0.213$). Additionally, CalTrans noted that samples were collected by grab method and may not accurately represent the true EMC or pollutant removal.

In Mallin et al. (2002), concentrations of fecal coliforms were measured in monthly influent and effluent data collected for a 29-month period from three retention ponds in Wilmington, North Carolina. Mean fecal coliform concentrations in the effluent of the three ponds were 70, 43 and 85 MPN/100mL (Table 5-4).

The Demonstration Urban Stormwater Treatment (DUST) Marsh at Coyote Hills Regional Park in Fremont (Alameda County), California was designed as a prototype system and research facility to study wetland creation for stormwater treatment in the San Francisco Bay Area. The entire project site is approximately 22 hectares and it receives urban runoff from a 1,178 hectare area within the city of Fremont, California. Runoff water enters the initial debris basin and is divided among two parallel flow systems (a lagoon and a pond system) that may be operated independently (ABAG, 2000). The two systems discharge into a common third system (a marsh system). Results for the DUST marsh system are presented in Table 5-4.

Davies and Bavor (2000) conducted a study on the Woodcroft Estate retention pond approximately 40 km northwest of Sydney, Australia. Water samples were collected from the inflow and outflow locations on a weekly basis from July through December 1998. The retention pond displayed a mean bacterial removal efficiency for thermtolerant coliforms and *Enterococcus* of -2.5 and 23% (-0.01and 0.11log), respectively (Table 5-4). The authors concluded that the poor bacterial removal efficiency resulted from the inability of the pond system to retain fine clay particles (<2 μm) to which bacteria were predominately absorbed.

A study was conducted in the Toronto, Canada area from 1994 to 1999 to evaluate the efficiency of three retention ponds during various climatic conditions (CWP, 2000). Monitoring included collecting grab samples at inflow and outflow locations in each retention pond during 20 storm events and analyzing for fecal coliforms. Removal was variable, ranging from 83 to -3 % (0.77 to -0.01 log) in the three ponds (Table 5-4).

Table 5-4. Retention Pond Removal Efficiencies for Microorganisms in Stormwater.

Study	Parameter	Mean EMC (unless otherwise noted)		% Removal	Log Removal	Significance ($p < 0.05$)
		Influent (MPN/100ml)	Effluent (MPN/100ml)			
CalTrans (2004)	Fecal Coliform	11,700	100	99	2.0	No
Mallin et al. (2002)	Fecal Coliforms (3 ponds)	488	70	85.6	0.84	Yes
		97	43	55.7	0.35	Yes
		74	85	-14.9	-0.06	No
ABAG (2000)	Fecal Coliforms	6,937	2,516	64	0.44	No
Davies and Bavor (2000)	Thermotolerant Coliforms	7,900	8,100	-2.5	-0.01	NR
	Enterococi	1,200	920	23	0.11	NR
CWP (2000)	Fecal Coliforms (3 ponds)	10,465	1,779	83	0.77	NR
		2,772	2,858	-3	-0.01	NR
		NR	783	NR	NR	NR
Kurz (1999) 5-day Shallow/Deep Pond	Fecal Coliforms	NR	NR	98.2/88.5[1]	1.62/1.82	Yes
	MS2 Coliphage[2]	NR	NR	93.9/98.6[1]	1.91/2.12	Yes
	Flourescent Latex Beads (*Cryptosporidium* surrogate)[3]	NR	NR	99.5/99.0[1]	2.48/2.07	Yes
Kurz (1999) 14-day Shallow/Deep Pond	Fecal Coliforms	NR	NR	76.4/69.2[1]	1.67/1.01	Yes
	MS2 Coliphage[2]	NR	NR	88.9/94.7[1]	1.01/1.56	Yes
	Flourescent Latex Beads (*Cryptosporidium* surrogate)[3]	NR	NR	99.1/99.5[1]	1.91/2.33	Yes
City of Austin (1997)	Fecal Coliforms	NR	1324	98	1.7	NR
	Fecal Streptococci	NR	1265	96	1.4	NR
Borden et al. (1996)	Fecal Coliforms (2 ponds)	NR	NR	48	0.28	Yes
		NR	NR	-5	-0.02	Yes

[1] % removal efficiencies were calculated using loads, as opposed to concentrations.
[2] PFU/mL
[3] beads/mL
NR = Not Reported

Kurz (1999) reported monitoring results from two retention ponds located at the Southwest Florida Water Management District's Tampa Service Office. The retention ponds were each 0.46 hectares (1.15 acres) in surface area, but differed in average depth (1.0 and 2.7 meters). Stormwater was extracted from the Tampa Bypass Canal into each pond on two separate occasions to simulate two different storm event treatment episodes. In the first simulated storm event, outflow was adjusted to increase the residence time of runoff within each pond to 14 days. In the second event, outflow was increased to achieve a residence time of five days. During both simulated storm events, MS2 coliphage (enteric virus surrogate) and fluorescent latex beads (*Cryptosporidium* spp. surrogate) were blended with the influent. Concentrations were such that each could be analyzed for a relatively small outflow sample volume. The results for the 5-day and 14-day retention ponds study are summarized in Table 5-4. It should be noted that the mean log removals are based on concentrations and the load removal efficiencies are based on loads. The retention ponds were able to achieve significant ($p<0.05$) reductions in total and fecal coliform, coliphage and fluorescent bead concentrations.

The City of Austin (1997) conducted a monitoring study on a retention pond in the Austin, Texas area from 1994 to 1996. A total of 20 samples were collected from the pond's inlets and outlets over the two year period. Results indicated a high level of fecal coliform and fecal streptococci removal (Table 5-4).

Borden et al. (1996) conducted an intensive monitoring study to document the pollutant removal performance of two retention ponds in the High Point, North Carolina area. Davis Pond and Piedmont Ponds were each sampled for one year, during which 22 and 25 storms were sampled for each pond, respectively. The annual mean fecal coliform reduction was reported as 48% (0.28 log) for the Davis Pond and -5%(-0.021 log) for Piedmont Pond (Table 5-4). Borden et al. (1997) speculated that longer drawdown time in the Davis Pond may have facilitated the settling of sediments and associated bacteria, resulting in the higher removal of fecal coliform than the Piedmont pond. Additionally, removal of coarse sediments by the upstream pond may have reduced particle settlement and pollutant removal in the Piedmont pond.

5.3.2 Application in Urbanized Areas and Costs

Retention ponds have been used extensively throughout urbanized areas in the U.S. for many years. However, similar to detention basins, it is difficult to use retention ponds in urban areas because of the area required to construct the pond, usually two to three percent of the contributing drainage area (based on design requirements). Specifically, in humid climates a 10-hectare drainage area is generally required, and larger drainage areas are needed in semi-arid and arid climates (CWP, 2001). Additionally, sufficient precipitation and year round flow in the drainage area are also needed.

Retention ponds are relatively inexpensive structural control technologies, however, there is a wide range of costs associated with their construction. General construction costs are estimated between \$10 and \$37/m^3 (Brown and Schueler, 1997; Wanye County, 2001). Design and permitting are estimated to be roughly 30 percent of the construction cost.

Maintenance activities are similar to those described for detention basins, although retention ponds require additional maintenance to avoid stratification that could cause solids (and bacteria) to become resuspended (CalTrans, 2004). Annual maintenance costs have been estimated in a variety of studies. Schueler (1992) estimated that maintenance costs for retention ponds were roughly three to five percent of construction costs. CalTrans (2004) estimated that annual maintenance costs for a retrofit retention pond were \$17,632, including inspections, maintenance, vector control, administrative duties, and materials. Maintenance costs were also estimated by Wayne County (2001) at $10,000 per year and include activities such as sediment removal.

5.4 Infiltration Basins and Trenches

An infiltration basin is a shallow stormwater runoff impoundment designed to receive, hold and eventually infiltrate stormwater runoff into the ground. Infiltration basins use the ability of the soil to naturally filter runoff as the mechanism for pollutant removal. The runoff is held until it permeates the soil after a period of days and ultimately moves into groundwater.

Infiltration trenches are typically much smaller than basins, but treat stormwater mainly through the same process, soil filtration. Trenches are constructed by excavating, lining with filter fabric and filling with crushed stone, ranging from 0.91 to 3.66 meters (3 to 12 feet) in depth (U.S. EPA, 1999a). Runoff is stored in the void space of a stone aggregate in the trench and gradually infiltrates into the subsurface.

5.4.1 Removal Efficiencies

No case studies were found during the literature review on the efficiency of infiltration basins and trenches to remove bacterial indicators and/or pathogens in stormwater. This lack of information is likely due to the difficulty involved in collecting samples in the "effluent" that has infiltrated into the underlying soil. The U.S. EPA's Preliminary Data Summary of Urban Stormwater Best Management Practices states that infiltration systems can be considered "100 percent effective" at removing indicator bacteria in infiltrated stormwater (U.S. EPA, 1999b). However, pollutant removal efficiency is directly associated with how well the BMP is designed and maintained. The risk of groundwater contamination may also be a concern in areas with sandy soils with sallow aquifers, especially for viruses (Pitt et al., 1994)

5.4.2 Application in Urbanized Areas and Costs

Infiltration basins and trenches have not been implemented in the urban landscape to the same extent as other the structural stormwater treatment controls. A variety of factors (i.e., design criteria, inspection and maintenance, and costs) are likely responsible for the lack of implementation. Infiltration trenches and basins are typically applicable for drainage areas under two hectares. Additionally, these BMPs perform best in permeable soils (Hilding, 1996).

Infiltration trenches are frequently applied to sites with restricted land availability, and are typically constructed in warmer, more humid regions, although recent studies conducted on infiltration trenches in Maryland and New Jersey have shown them to be suitable in colder climates as long as the design prevents icing (U.S. EPA, 1999a). It is also important to point out that many agencies believe that infiltration trenches and basins are not generally applicable in ultra urban areas due to the possibility for infiltrated runoff to interfere with existing infrastructure and the low permeability of most urban soils (CWP, 2001).

Although costs vary, infiltration basins are typically cost-effective to construct because of minimal amount of infrastructure and land needed. Infiltration basins typically only consume about 2 to 3% of the site draining to them. Based on CalTrans (2004), the average cost to construct an infiltration basin in 1999 dollars was $369/m3. In contrast, infiltration trenches are slightly more expensive, based on the cost per area treated relative to other structural treatment controls (CWP, 2001). Construction costs included clearing, excavation, lining of the trench with filter fabric and rock, and placement of a vegetated buffer strip. The Southeastern Wisconsin Regional Planning Commission (1991) estimated construction costs for a 68 m^3 infiltration trench to be between $8,000 and $19,000 (1993 dollars). Smaller trenches were estimated to range between $3,000 and $8,500 (SWRPC, 1991). CalTrans (2004) estimated mean adjusted construction costs (1999 dollars) at $733/m3.

A major disadvantage to implementing infiltration basins and trenches is the routine inspection and maintenance needed to reduce the potential for the system to clog. This increases the maintenance costs of infiltration basins to an estimated at 5-10% of construction costs, well above most other treatment controls. CalTrans (2004) estimated the annual operation and maintenance costs of a infiltration basin and trench to be roughly $3,000 and $2,639, respectively, including inspections, maintenance, vector control, administration, and direct costs.

5.5 Vegetated Swales

A vegetated swale, also known as a grassed channel, dry swale, retention swale, biofilter, grassy or grassed swale, is a natural or constructed wide, shallow channel with side slopes and base covered with vegetation. Vegetated swales are employed to capture particulate pollutants, promote infiltration, and decrease the rate of runoff. Runoff is primarily treated by filtering through the vegetation in the channel and/or through infiltration into a subsoil matrix and underlying soils (CASQA, 2003). Vegetated swales can be used instead of curbs, gutters and storm sewer systems, thereby acting as one component of the storm drain system.

5.5.1 Removal Efficiencies

The following section briefly summarizes readily available data on the efficiency of vegetated swales in removing microorganisms in stormwater. Direct comparison of these data are not possible since watershed area, sampling methodology, number of storms monitored, swale design, and removal efficiency techniques differed for each study. However, the information presented from these studies may provide a general range of pollutant reduction that can be expected from swales.[12] The studies are presented in chronological order, beginning with the most recent.

CalTrans (2004) reported on the pollutant removal efficiencies of six vegetated swales in Southern California. Each swale was located in a highway right-of-way and treated highway runoff. Influent and effluent grab samples were collected during storms and analyzed for chemical and microbial indicators. Mean fecal coliform EMCs for the six swales were 12,300 MPN/100mL and 16,000 MPN/100mL for the influent and effluent, respectively (Table 5-5). These data equate to a -30%(-0.11 log) removal ($p = 0.707$ based on ANOVA). However, it is important to note that coliform samples were collected by the grab method and may not

---

[12] Although influent and effluent concentrations are provided in this summary, it is important to note that the better indicator of removal efficiency for BMPs that infiltrate water is pollutant loads.

accurately reflect removal.

Jones and Langan (1996) studied the pollutant removal efficiencies of a variety of stormwater treatment controls constructed in the Great Bay watershed in New Hampshire. A total of five vegetated swales that receive runoff from commercial land uses were sampled during the study. Concentrations of fecal coliforms and *E. coli* were analyzed in swale influent and effluent five times during the study. As illustrated in Table 5-5, effluent concentrations were generally greater than influent concentrations in all swales sampled. Goldberg (1993) conducted a pollutant removal efficiency study in the greater Seattle area on a vegetated swale located in a right-of-way separating backyards in a residential setting. Sampling was conducted during eight storm events between 1991 and 1993. Results showed an estimated fecal coliform removal of –264% (–0.56 log) with a corresponding average inflow EMC of 3,725 CFU/100mL (Table 5-5). The internal sources that were attributed to the increase in fecal coliforms included pets that routinely defecated in the swale and a beaver colony with burrows in the slide slopes (Goldberg, 1993).

Table 5-5. Vegetated Swale Removal Efficiencies for Microorganisms in Stormwater.

Study	Parameter	Mean EMC (unless otherwise noted) Influent (MPN/100ml)	Mean EMC (unless otherwise noted) Effluent (MPN/100ml)	% Removal	Log Removal	Significance ($p < 0.05$)
CalTrans (2004)	Fecal Coliforms (6 swales)	12,300[1]	16,000[1]	-30.1	-0.11	No
Jones and Langan (1996)	Fecal Coliforms (5 swales)	210	276	-31.4	-0.12	NR
		100	177	-77.0	-0.25	NR
		2	15	-650	-0.88	NR
		367	158	57.0	0.37	NR
		5	6	-20	-0.08	NR
Jones and Langan (1996)	*E. coli* (5 swales)	154	125	18.8	0.09	NR
		85	131	-54.1	-0.19	NR
		1	14	-1300	-1.15	NR
		186	119	36.0	0.19	NR
		6	2	66.7	0.48	NR
Goldberg (1993)	Fecal Coliforms	3,725	13,559	-264	-0.56	NR
Seattle Metro and Washington Department of Ecology (1992)	Fecal Coliforms	NR	NR	Negative	NR	NR

[1] Mean EMC for 6 swales

NR = Not Reported

The City of Mountlake Terrace, Washington, constructed a 70-meter long vegetated swale and studied pollutant removal capability (SMWE, 1992). The swale served a 6.3-hectare drainage area with 47% imperviousness composed of residential areas, parks and a heavily traveled road. During the second part of the study half of the swale was bypassed by a pipe, reducing its effective length to 30.5 meters and residence time to about five minutes. During the summer and fall of 1991, runoff from the 70-meter configuration was monitored during six storm events. In addition, six flow-weighted composite samples were gathered from the 30.5-meter configuration during the fall and winter of 1992. Seattle Metro and Washington Ecology (1992) reported that fecal coliform concentrations in the swale effluent were greater than influent concentrations, making the pollutant removal efficiency negative (Table 5-5). The increase in bacteria in swale effluent was attributed to the presence of pet waste and the possible re-growth of bacteria within the vegetated swale itself (SMWE, 1992).

5.5.2 Application in Urbanized Areas and Costs

The use of vegetated swales as stormwater treatment BMPs in urban areas has grown steadily over the last 10 years. A variety of factors (i.e., design criteria, inspection and maintenance, and costs) are likely responsible for the increase in implementation. Swales can be sited in most regions of the U.S., though in arid and semi-arid regions they may require regular irrigating during the dry months. Due to their linear shape, vegetated swales are generally applicable to sites receiving highway or residential road stormwater runoff. However, it should be noted that vegetated swales can only drain small areas (preferably less than 2 hectares) effectively, since high flow velocities generally lead to bypass of treatment (CWP, 2001). Additionally, swales should only be constructed on sites with slopes less than four percent.

Compared to other BMPs, vegetated swales have relatively low construction costs, although costs vary depending on size of the swale. Additionally, swales are cost effective alternatives to traditional stormwater conveyance systems (e.g., curb and gutter). Schueler (1987) reported that costs may range from $16-30 per linear meter for the construction of a swale with a 4.5-meter wide channel (top width). Costs have also been estimated to range from $28-164 per linear meter depending on the width and depth of the system (SWRPC, 1991).

The majority of maintenance associated with vegetated swales relates to maintenance of the vegetated cover. Mowing and rototilling should be conducted on an as needed basis. Annual maintenance costs for swales have been estimated to be roughly $1.90 per linear meter for a 0.5-meter deep swale (SWRPC, 1991).

5.6 Stormwater Wetlands

A stormwater wetland, also known as a constructed wetland, wetland basin, shallow marsh, extended detention wetland, wetland system, pocket wetland and gravel based wetland, is similar to a retention pond (with a permanent pool of water) but has more than 50% of its surface covered by emergent wetland vegetation. Additionally, stormwater wetlands are much shallower than retention ponds, typically averaging 0.3 to 0.5 meters in depth. Wetlands also have plants suited to growing in saturated soils. Microorganism removal from the water column occurs mainly through sedimentation, predation and inactivation via sunlight (Davies and Bavor, 2000).

Stormwater wetlands are different from natural wetlands in that they are solely designed to treat stormwater runoff and therefore have fewer species of flora and fauna. Stormwater wetlands are a widely used structural treatment control, although their use is restricted in highly urban areas and to some degree in arid climates (CWP, 2001).

5.6.1 Removal Efficiencies

The following section briefly summarizes readily available data on the efficiency of stormwater wetlands in removing microorganisms. Direct comparison of these data are not possible since watershed area, sampling methodology, number of storms monitored, swale design, and removal efficiency techniques differed for each study. However, the information presented from these studies does provide a general range of pollutant reduction that can be expected from stormwater wetlands. The studies are presented in chronological order, beginning with the most recent.

Hinds et al. (2004) investigated fecal coliform removal efficiencies of two constructed wetlands at the Olentangy River Wetland Research Park located at Ohio State University in Columbus Ohio. Samples were collected on a monthly basis from October 2000 to June 2001, excluding months when the wetlands were frozen (December 2000 through February 2001). Results showed that although fecal coliform concentrations differed among months, the concentration levels were always lower at the outlet than at the inlet location. Several factors in wetland systems allow for the treatment of bacteria in these systems including the amount of plant coverage, hydraulic retention time and the settling velocities of microorganisms. A summary of the results are presented in Table 5-6.

Table 5-6. Stormwater Wetland Removal Efficiencies for Microorganisms in Stormwater.

Study	Parameter	Mean EMC (unless otherwise noted)		% Removal	Log Removal	Significance ($p < 0.05$)
		Influent (MPN/100ml)	Effluent (MPN/100ml)			
Hinds et al. (2004) Wetland #1	Fecal Coliforms	35	6.4	81.7	0.74	NR
Hinds et al. (2004) Wetland #2	Fecal Coliforms	32.6	8.4	74.2	0.59	NR
City of Laguna Niguel and County of Orange (2004)	*Enteroccocus*	NR	NR	60.4 to 99.6	0.4 to 2.4	No
Bruce (2003) Wetland #1	Fecal Coliforms	320[1]	8[1]	97	1.6	NR
	Enterococcus	544[1]	29[1]	65	1.3	NR
Bruce (2003) Wetland #2	Fecal Coliforms	300[1]	39[1]	87	0.9	NR
	Enterococcus	1,460[1]	361[1]	75	0.6	NR
Davies and Bavor (2000)	Thermotolerant Coliforms	17,000	3,600	78.8	0.67	Yes
	Enterococcus	6,100	900	85.2	0.82	Yes

[1] Event median concentration
[2] PFU/mL
[3] cysts/100L
[4] oocusts/100L
NR = Not Reported

The City of Laguna Niguel and the County of Orange (2004), California constructed a wetland system to address elevated levels of fecal coliforms in their stormwater drainage area. The wetland system consisted of three wetlands that concentrated mainly on treating dry-weather runoff from the surrounding drainage area (~2.3 km2). Outlet samples were collected prior to wetland system construction and compared to samples collected after the wetland system was put online using a one-way analysis of variance (ANOVA). No significant difference of the *Enterococcus* concentration geometric means was determined, although removal efficiencies for Enterooccus ranged from 60.4-99.6% (0.4-2.4 log).

Bruce (2003) investigated two constructed wetlands located in the vicinity of Pendray Dairy Farms in North Saanich, British Colombia. Sampling was conducted from October 2000 to May 2002. Results indicated that both wetlands reduced the concentrations of fecal coliform and *Enterococcus* (Table 5-6), although no analyses were performed to determine if the reductions were statistically significant.

Davies and Bavor (2000) conducted a study on a 0.45-hectare constructed wetland on the Plumpton Estate residential development about 40 km northwest of Sydney, Australia. Water samples were collected from each of the inflow and outflow locations on a weekly basis from July through December 1998 and thermotolerant coliforms and *Enterococcus* were enumerated. Outflow concentrations of all indicator organisms were generally lower than inflow concentrations (Table 5-6). The authors speculate that the wetland's efficiency in removing indicator bacteria was related to plant material in the wetland impeding flow, which improved sedimentation and the removal of bacteria adsorbed to fine particles.

5.6.2 Application in Urbanized Areas and Costs

Stormwater wetlands are a popular BMP implemented in urbanized areas in a variety of climates. However, similar to detention basins and retention ponds, stormwater wetlands are difficult to utilize in urbanized areas due in part to the land area constraints. In humid climates, approximately 10 hectares of drainage area are necessary to maintain a pool, and larger drainage areas may be required in more arid climates (CWP, 2001). Additionally, stormwater wetlands can only be sited in locations with a sufficient flow of water and suitable soils. There must be adequate baseflow from the watershed or groundwater to maintain a permanent pool and to sustain plant growth.

General construction costs for stormwater wetlands have been estimated by a variety of sources. Construction costs largely depend upon the amount of excavation required at the site and plant selection. The Center for Watershed Protection (2001) estimated that construction costs are between \$12/m3 and \$46/m^3, disregarding the price of land. Alternatively, the U.S. EPA (1999a) estimated that the construction costs for a stormwater wetland with a sediment forebay range from \$65,000-137,500 per hectare of wetland.

Properly designed and well-maintained stormwater wetlands can perform efficiently for at least 20 years (CWP, 2001). However, maintenance activities must be routinely performed and must begin during construction of the system. Wetlands should be inspected for invasive vegetation semi-annually and annually inspected to ensure that inlet and outlet devices are free of debris and operational. Wetland plants smothered by sediment should be removed periodically. Sediment removal from the wetland is typically required every 20 to 50 years, when the pool volume becomes reduced substantially (CWP, 2001). The mean annual routine maintenance cost for wetlands is generally estimated at two to five percent of the construction cost (CWP, 2001). Wayne County (2001) estimated maintenance costs between \$1,500 and \$2,700 per wetland hectare.

5.7 Sand Filters

The standard sand filter design is a two-chambered or three-chambered system. The first chamber is used for the settling of particles and the second chamber is a filter bed composed of sand or a similar filtering matrix. Stormwater runoff moves through the first chamber and large particles settle to the bottom. Fine particles and additional pollutants are then filtered out as runoff passes through the filtering matrix in the second (and third, if applicable) chamber.

Although there are many adaptations, there are three basic types of sand filters used to treat stormwater. These include the Austin, Washington, D.C. and Delaware sand filters (U.S. EPA 1999a). The Austin sand filter is composed of a sedimentation basin and a sand matrix filter. Similar to the Austin, the Delaware sand filter is also a two-chambered system, consisting of a sedimentation chamber and a filtration chamber. Runoff enters the sedimentation basin through a grated cover before it moves into the filtration chamber, which is composed of a thick layer of sand. The Washington, D.C. sand filter is similar to the Austin and Delaware, however, it is comprised of three underground chambers, of which one is for sedimentation, one for filtration and one for discharge. Pollutant removal by sand filters generally occurs by filtering pollutants through the filtration matrix and through the settling of solids in the sedimentation basin (CalTrans, 2004).

5.7.1 Removal Efficiencies

The following section briefly summarizes readily available data on the efficiency of sand filters in removing microorganisms in stormwater. Direct comparison of these data are not possible since watershed area, sampling methodology, number of storms monitored, swale design, and removal efficiency techniques differed for each study. However, the information presented from these studies does provide a general range of pollutant reduction that can be expected. The studies are presented in chronological order, beginning with the most recent.

CalTrans (2004) sited, constructed and monitored six Austin sand filter systems in Southern California. The filters were constructed in maintenance yards or park-and-ride locations that allowed for a restricted work area. Influent and effluent grab samples were collected during storms and analyzed for chemical and bacterial constituents. Mean fecal coliform EMCs for the influent and effluent samples are presented in Table 5-7. CalTrans (2004) reported a 72% (0.55 log) removal although influent and effluent concentrations were not significantly different, based on ANOVA. CalTrans noted that coliform samples were collected by grab method and may not accurately reflect removal.

CalTrans (2004) also sited, constructed and monitored one Delaware sand filter in a maintenance yard in Southern California. Influent and effluent grab samples were collected during storms and analyzed for chemical and bacterial constituents. CalTrans (2004) reported a 79% (0.68 log) removal although influent and effluent concentrations were not significantly different, based on ANOVA (Table 5-7).

Kurz (1999) reported on the monitoring of a sand filter that received runoff from a 2.73-hectare light commercial/urban watershed in the City of Madeira Beach, Florida. Sampling was conducted between September 1995 and November 1996. Runoff was treated as water moved through one of three sand filter chambers, each containing roughly one meter of gravel and sand. The water then moved through a gravel underdrain and porous drainage pipe, and was released to a canal. During each sampling event, roughly 5,160 liters of residual runoff captured in a holding tank was transferred into one of the three filters. MS2 coliphage (enteric viral indicator) and fluorescent latex beads (*Cryptosporidium* spp. surrogate) were blended with the raw runoff at concentrations that would allow enumeration to occur in a relatively small outflow sample volume. Additionally, total and fecal coliforms were enumerated. The results are summarized in Table 5-7. In most of the trials, the sand filters were able to achieve significant ($p<0.05$) reductions in fecal coliform and coliphage concentrations.

Table 5-7. Sanf Filter Removal Efficiencies for Microorganisms in Stormwater.

Study	Parameter	Mean EMC (unless otherwise noted)		% Removal	Log Removal	Significance ($p < 0.05$)
		Influent (MPN/100ml)	Effluent (MPN/100ml)			
CalTrans (2004) Six Austin Sand Filters	Fecal Coliforms	5,800	1,600	72.4	0.56	No
CalTrans (2004) Delaware Sand Filter	Fecal Coliforms	5,800	1,200	79.3	0.68	No
Kurtz (1999)	Total Coliforms	NR	NR	NR	0.88	No
	Fecal Coliforms	NR	NR	NR	1.01	Yes
	MS2 Coliphage[1]	NR	NR	NR	2.02	Yes
	Fluorescent latex beads (*Cryptosporidium* surrogate)[2]	NR	NR	NR	3.57	No
City of Austin (1997)	Fecal Coliforms	NR	18,528	-85.0	-0.27	NR
	Fecal Streptococcus	NR	2,573	69.0	0.51	NR

[1] PFU/mL
[2] beads/mL
NR = Not Reported

Another monitoring study on a sand filter system was conducted by the City of Austin (1997). The sand filter was referred to as the Barton Creek Plaza and received runoff from just under three acres of a shopping center parking lot in Austin, Texas. Nine storm events were sampled from 1996 to 1997 for fecal coliform and fecal streptococcus. Although removal efficiencies for fecal streptococcus were positive, mean fecal coliform outflow concentrations were substantially greater than inflow (Table 5-7). It was concluded that the removal of bacteria was quite variable and evidently the warm, dark and damp environment of the sand filter sometimes served as a source of fecal coliform bacteria.

5.7.2 Application in Urbanized Areas and Costs

Sand filters are popular stormwater treatment controls used throughout the U.S.

The Washington, D.C and Delaware sand filters are designed to treat runoff from highly impervious areas up to 0.40 and 2 hectares in size, respectively. They are best suited to very impervious areas where space is restricted, because they are belowground systems. The Austin sand filter system is best suited for larger drainage areas (< 20 hectares) that have a mixture of impervious and pervious surfaces (U.S. EPA, 1999a).

The construction costs for Washington D.C. and Austin sand filter systems serving a drainage area under 0.40 hectare are estimated to be approximately \$6,600 to \$11,000 and \$18,500, respectively (1997 dollars) (U.S. EPA, 1999a). The Delaware sand filter system costs are similar to the Washington, D.C. sand filter system, except that excavation costs will be less due to the shallow construction of this system (U.S. EPA, 1999a).

Sand filters require inspection immediately following storm events since they are prone to clogging. In addition, common maintenance includes the removal of trash and debris and substitution of the uppermost layers of sand or other filter media if clogged. Annual maintenance costs for sand filter systems are estimated at five percent of the initial construction cost (Schueler, 1992).

5.8 Proprietary Devices

Over the past two decades, a variety of proprietary devices have been developed to remove constituents readily found in stormwater. These devices differ from traditional stormwater treatment methods that were previously discussed in this chapter in that they are extremely compact and can be retrofitted into existing stormwater conveyance systems. This section provides a brief overview of the various categories of proprietary devices designed to treat stormwater and discusses the demonstrated pollutant removal efficiencies and costs (were available) of implementing and maintaining these devices.

It is important to note that although manufacturers' literature promoting these devices typically includes information regarding pollutant removal efficiencies, it is rarely accompanied by adequate available supporting documentation (Woodward-Clyde, 1998). Furthermore, most manufacturers' tests are performed under laboratory conditions and do not replicate conditions likely to be found in actual installations in the field (Stenstrom, 1998; Woodward-Clyde, 1998). Therefore, information from manufacturers was not included in this summary.

5.8.1 Gravity Separation Devices

Gravity separation units are generally designed to remove suspended solids without land requirements associated with traditional technologies. Removal efficiency of gravity separation devices is a function of the detention time and settling velocity of the particles of the influent (Brueske, 2000). Proprietary devices that use gravity separation as their primary treatment mechanism include the Stormceptor® and BaySaver® systems.

The project team found one study that assessed the efficiency of gravity separation devices to remove microorganisms in stormwater. In 2002, Clausen et al. studied the efficiency of a Stormceptor® system to remove a variety of constituents, including fecal coliforms. Results indicated that the average fecal coliform concentrations in influent and effluent (n=7) were not statistically different (p=0.05).

Additionally, one review (SCVURPPP, 2000) reviewed water quality engineering fundamentals to assess the potential efficiency of gravity separators and concluded that if adequately sized and maintained, these devices may be able to provide removal of grit and coarse particles. However, they are not expected to be effective for removing dissolved constituents or those attached to fine particles, such as microorganisms.

5.8.2 Swirl Concentration Devices

Swirl concentration devices have been used since the early 1960s to treat combined sewer overflows (CSOs) and adapted for use in combined sewer systems in the U.S. in the 1970s. Swirl concentrators use vortex separation as a mechanism for removing solids from stormwater. Flow entry into the device is designed to cause a swirling motion around the removal chamber. These devices rely on the secondary flow current induced by the vortex to separate and draw solids downwards to a gutter while allowing "clear" water to overflow at the top of the separation chamber. Proprietary devices that use swirl concentration as their primary treatment mechanism include the Vortechs™ and Downstream Defender™ systems.

Only a few studies have evaluated the effectiveness of swirl concentration devices in removing microorganisms in stormwater. West et al. (2001) and Clausen et al. (2002) studied the effectiveness of a Vortechs™ stormwater treatment system and found that the devices had negligible effects on fecal coliform concentrations. Clausen et al. (2002) also reported that Downstream Defender™ system had similar results. Additionally, similar to gravity separators, engineer fundamentals suggest that swirl concentration devices may be effective at removing coarse particles, but are unlikely to capture fine particles and will be ineffective for dissolved pollutants (SCVURPPP, 2000). Based on these results, it is unlikely that swirl concentrators have a high efficiency for removing microorganisms in stormwater.

5.8.3 Filtration Devices

There are many different types of filters used in stormwater treatment devices. They can be grouped into three basic types: strainers, granular media, and adsorptive filters. In strainers (which include drain inlet liners) the primary mechanism of removal is physical straining through a filter mesh. In granular media filters, sand is the most common media used in the filter beds. The mechanisms for removal in filters of this type are more complex than just physical straining. Adsorptive filters are filters that, in addition to straining and granular filtration, make use of chemical adsorption as an additional removal mechanism. Stormfilter™ is an example of a proprietary device that uses filtration as its primary treatment mechanism.

Although no studies were found that evaluated the efficiency of proprietary filtration devices to remove microorganisms in stormwater, non-proprietary sand filters have shown moderate removal rates for coliform bacteria (see Section 5.7). However, stormwater filtration systems do require bypass to prevent overflow and not compromise the hydraulic integrity of the stormdrain. Additionally, a sedimentation chamber is typically required as part of the filter system to reduce clogging, which will reduce efficiency.

5.8.4 Application in Urbanized Areas and Costs

Proprietary stormwater treatment devices are designed for implementation in highly urbanized areas. Therefore, site restrictions based on limitations in land area available are less than those for non-proprietary structural controls (e.g., detention basins, retention ponds). However, due to flow and volume considerations, proprietary devices typically achieve higher pollutant removal efficiency when treating relatively small areas (<5 hectares).

Construction costs for proprietary treatment devices vary, depending on the type of device and whether it is being installed as part of new construction or a retrofit project. Generally, it is cheaper to install a device during new construction. Example equipment and construction costs are presented in Table 5-8.

Proprietary treatment control devices are typically designed to retain captured pollutants over multiple rain events. To ensure proper functioning, these devices should be inspected, and floatable materials and accumulated sediment should be removed when devices are 50-85% full (CalTrans, 2004). Maintenance costs for proprietary devices were not available for this review, but will likely vary among devices.

Table 5-8. Estimated Equipment and Construction Costs for Selected Proprietary Treatment Control Devices.
(Clausen, J.C., Belanger, P., Board, S., Dietz, M., Phillips, R., Sonstrom, R. Stormwater Treatment Devices Section 319 Project: Project #99-07. Hartford, CT: Connecticut Department of Environmental Protection. 2002)

System	Construction Type	System Cost ($)	Design & Installation Cost ($)	Total Cost ($)
Catch Basin	New	2,500	600	3,100
Downstream	Retrofit	24,795	8,000	32,795
Stormceptor®	New	12,500	2,500	16,000
Stormtreat™	Retrofit	6,600	20,000	26,600
Vortechs™	New	16,500	8,000	25,304

5.9 Advanced Treatment and Disinfection

Although, disinfection of wastewater treatment plant effluent, industrial wastewater discharges and combined sewer overflows (CSOs) is an established practice, only a few disinfection facilities for discharges from separate storm sewer systems have been developed world-wide (Perdek et al., 2003). This section provides an overview of the various types of disinfection technologies and discusses the demonstrated pollutant removal efficiencies and costs of facilities designed to disinfect stormwater. It is not the intent of this section to provide a thorough discussion of disinfection processes.

5.9.1 Overview of Disinfection Technologies

The three common methods of disinfection in the U.S. are chlorination, ozonation, and ultraviolet (UV) disinfection. Chlorine, the most widely used disinfectant for municipal wastewater, destroys target organisms by oxidation of cellular material. Like chlorine, ozone is a strong oxidizing agent. It is an unstable gas that is generated by an electrical discharge through dry air or pure oxygen. Ultra Violet (UV) irradiation, generated by an electrical discharge through mercury vapor, penetrates the genetic material of microorganisms and retards their ability to reproduce. There is no perfect disinfectant. However, there are certain characteristics to look for when choosing a suitable disinfectant for a treatment facility:

- ability to penetrate and destroy infectious agents under normal operating conditions;
- lack of characteristics that could be hazardous to people and the environment before or during disinfection;
- safe and easy handling, storage, and shipping;
- absence of toxic residuals and mutagenic or carcinogenic compounds after disinfection; and,
- affordable capital and operation maintenance (O&M) costs.

5.9.1.1 Ozonation

Ozone is the strongest and fastest-acting oxidant of all classical disinfecting agents used for water disinfection today. However, organic and inorganic impurities, chemical oxygen demand (COD), pH, temperature and suspended solids in waters have a significant impact on ozone disinfection efficiency (Perdek et al., 2003). Four basic components comprise an ozone treatment system: a gas feed system, an ozone generator, an ozone contactor, and an off-gas destruction system. The gas feed system provides a clean and dry source of oxygen to the generator. The ozone contractor then conveys the ozone-rich gas into the water for treatment and allows contact time for disinfection to occur. Finally, a process involving off-gas destruction is necessary since the level of ozone present in the off-gas is toxic (CalTrans, 2004).

5.9.1.2 Chlorination/Declorination

Chlorine or its derivatives are the most commonly applied disinfectants in the U.S., although its use has come under scrutiny over the last 20 years, due to concerns about suspected carcinogen byproducts (i.e., chloroform and trihalomethanes). Chlorine is readily available in many forms, which include chlorine gas, hypochlorite solutions, and other chlorine compounds in solid or liquid form. Chlorine is effective at inactivating bacteria and viruses, although its effectiveness against protozoa is questionable. Most states will not allow the use of chlorination alone because of residual chlorine concentrations in water and their likely adverse effect on receiving waters and aquatic species. Therefore, dechlorination is typically required to remove the excess chlorine. If not dechlorinated, chlorine residual can persist in the effluent for hours.

5.9.1.3 Ultraviolet Light Irradiation

Ultraviolet (UV) light disinfects water by damaging microorganism DNA to prevent multiplication and infection. The UV disinfection works by utilizing germicidal lamps within a pressure vessel or underneath a water channel. Microorganisms in the water moving past the UV lamps are subjected to lethal amounts of UV energy. When used to treat stormwater runoff, the UV disinfection system is typically an in-line system located downstream of another treatment system such as a sand filter, sedimentation basin or multiple chamber treatment train (CalTrans, 2004).

5.9.2 Stormwater Advanced Treatment Case Studies

The following sections briefly summarize information on facilities designed to treat dry-weather flows from separate storm sewer systems. Treatment process information, costs and data on removal efficiencies is presented where available.

5.9.2.1 Moonlight Beach Urban Runoff Treatment Facility (Encinitas, California)

In response to numerous postings and closures of Moonlight Beach due to exceedances of ambient water quality criteria (AWQC) for indicator bacteria, the City of Encinitas completed the construction of a dry-weather urban runoff treatment facility in 2002. The facility is designed to treat dry-weather flow rates up to 150 gallons per minute (GPM) from Cottonwood Creek, which drains a 820 hectare highly urbanized watershed (Rasmus and Weldon, 2003). The facility construction occurred after several years of upstream BMP implementation and heavy enforcement of urban runoff regulations did not reduce the concentrations of microbial indicators below AWQC. City staff investigated pollutant sources upstream of Moonlight Beach and hundreds of notices of violations for urban runoff pollution infractions were issued to restaurants, gasoline stations, automotive repair shops, and other businesses. Additionally, structural (i.e., washing-containment facilities, detention basins and filters on new development, and drainage facilities) and nonstructural BMPs (i.e., good housekeeping practices, scheduled street sweeping, routine storm drain facility cleaning, and employee training programs) were employed (Rasmus and Weldon, 2003).

Prior to construction of the facility, the City of Encinitas had to address a variety of issues, including location, potential impacts to the creek and the type of disinfection method they would employ. Regarding location, initial consideration was given to diverting the dry-weather flow to a local wastewater treatment facility, however, costs of required pumping, conveyance and treatment facilities made this option unfeasible (Rasmus and Weldon, 2003). The city alternatively elected to place the facility within the existing footprint of a sewer pump station that is one block from the beach. This location was limited to a maximum footprint of 3 x 8 meters.

Potential impacts to the creek also had to be addressed. To maintain the biological integrity of the creek and perennial flow, the city estimated that 15% of the creek flow had to bypass the treatment facility. Additionally, no alternations to the natural streambed could be made. Therefore, the facility's inlet and outlet had to be located within an existing box culvert.

For the disinfection process, the City of Encinitas initially considered three types of methods; traditional chlorine, ozonation and ultra-violet (UV) irradiation. Concern about the potential danger of storing chlorine gas or liquid sodium hypochlorite removed the traditional chlorine disinfection from the list of alternatives. Ozone and UV disinfection were then evaluated in parallel. Ultimately, UV disinfection was selected because of its anticipated performance to meet AWQC, lower present-worth costs and lower cost of construction and maintenance. There were also two environmental advantages of UV disinfection. No chemical storage was needed and UV disinfection leaves no residual disinfectant in the treated water.

The Moonlight Light Beach Urban Runoff Treatment Facility has been highly effective at reducing concentrations of indicator bacteria in Cottonwood Creek and Moonlight beach. Removal efficiencies average 99.9% (log 3.0) for total coliforms, fecal coliforms and

Enterococcus. Total construction costs for the facility were roughly $480,000. Maintenance and operation costs were not readily available.

5.9.2.2 Santa Monica Urban Runoff Recycling Facility (Santa Monica, California)

As a way to reduce pollutants entering the Santa Monica Bay, the City of Santa Monica began to temporarily divert its dry-weather flows from two storm drainage areas to the City of Los Angeles's Hyperion Wastewater Treatment Plant in 1992. The temporary diversion occurred from 1992 to 2000 for six months per year (dry season) due to the treatment plant's inability to handle additional flow during winter months. In 2000, the diversion was replaced by the construction of the Santa Monica Urban Runoff Recycling Facility (SMURRF), which treats the diverted urban runoff and produces high quality water for reuse in landscape irrigation. The SMURRF is designed to treat a maximum dry-weather flow of 500,000 gallons per day (GPD) (Antich et al., 2003).

An evaluation was conducted of various water treatment processes to determine their ability to produce a reclaimed effluent suitable for landscape irrigation and toilet flushing, based on specific regulations. Similar to the Moonlight Beach treatment facility, a limited amount of space and on-going maintenance costs were issues that needed to considered. The preliminary treatment processes that were selected included bar screens and floatation units designed to remove large debris, oil and grease. Secondary treatment included microfiltration, which provides the same benefits as sand filtration, but has a smaller footprint and is able to handle a wider range of variable influent water quality (Antich et al., 2003).

Several disinfection processes were also evaluated, including ozonation, UV irradiation and sodium hypochlorite. The UV process was ultimately selected because of its small physical size requirements, minimal chemical handling, reduced environmental impact and present low-worth cost (Antich et al., 2003).

Because the two drainage areas no longer discharge to Santa Monica Bay during dry-weather, due to diversion, the City of Santa Monica has effectively reduced their dry-weather discharge of urban creek flows to zero. Additionally, an added benefit is that dry-weather flows are being recycled.

Capital costs for the SMURRF were $9 million, of which, $6.3 million was for treatment and $2.7 was for the distribution system. Ongoing maintenance and operation costs were not available.

5.9.3 Challenges in Implementing Advanced Stormwater Treatment

The disinfection of discharges solely from separate storm sewer systems has been limited to a few facilities designed to treat dry-weather flows. The main impetus of the advanced treatment is to reduce the potential public health concern associated with microbial indicators concentrations at beaches in Southern California. Although these facilities have been very successful in removing microorganisms in stormwater (3.0 log removal), challenges remain to the implementation of stormwater advanced treatment on a broader scale. These challenges are briefly described in the following sub-sections.

5.9.3.1 Suspended Solids Removal

An important factor that should be considered when evaluating the potential for advanced treatment of stormwater is the concentration of suspended solids in influent. Although highly variable, stormwater is typically characterized by high levels (>59 mg/L) of total suspended solids (Maestre and Pitt, 2005). To provide adequate removal of microorganisms in stormwater, a necessary first step in advanced treatment is removal of suspended solids. High levels of particulate matter in stormwater can provide a shield of protection for microbes exposed to a disinfecting agent (Sakamoto and Cairns, 1997). This typically occurs when microorganisms are occluded inside sediment particles. If suspended sediments are not removed, disinfection efficiency will likely be reduced. Particularly sensitive to suspended solids concentrations and particle occlusion is UV irradiation (Wojtenko et al., 2001). Suspended solids removal can occur through a pretreatment process such as sand filtration or sedimentation.

5.9.3.2 Capital and Ongoing Costs

Table 5-9 provides conceptual capital cost estimates for CSO disinfection options that are the most viable for stormwater (ozonation and UV irradiation). Capital costs were calculated based on the sizing necessary to allow for peak design flow and a 4-log reduction in fecal coliforms. Annual operating and maintenance costs were calculated based on 40 events per year with 15 million gallons requiring treatment during each event (Wojtenko and Stinson, 2003). In summary, capital and ongoing maintenance and operations costs are significantly greater than those for traditional stormwater BMPs discussed in Sections 6.2 to 6.7.

Table 5-9. Advanced Treatment and Disinfection Treatment Cost Estimates.
(Wojtenko, I., Stinson, M.K. Research Summary. CSO Disinfection Pilot Study: Spring Creek CSO Storage Facility Upgrade, EPA/600-R-02/077, U.S. EPA, Office of Research and Development, Cincinnati, OH. 2003)

Technology	Conceptual Level Disinfection Costs ($)[1]					
	Ozone			UV		
Peak Design Flow (cfs)	1,250	2,500	5,000	1,250	2,500	5,000
Capital Costs	19,221,000	24,560,000	30,539,000	48,052,000	67,272,000	87,774,000
Annual O&M Costs	534,000	587,000	657,000	248,000	497,000	992,000

[1]Costs are present worth in 2000 dollars

5.10 Summary

When determining the appropriate stormwater treatment control to use in a specific situation, a variety of factors should be considered, including safety, demonstrated effectiveness, process design fundamentals, construction and on-going costs, potential environmental/adverse effects, and the flow and characteristics of the stormwater that may be treated. However, possibly the most important consideration is the desired effluent quality. Although highly variable, concentrations of microbial indicators in discharges from separate storm sewer systems are generally one to two orders-of-magnitude greater than Ambient Water Quality Criteria (AWQC) for bacteria (see Table 2-4). Therefore, if it is determined that a specific AWQC for indicator bacteria is the effluent goal of a BMP or series of BMPs, the removal efficiency for microorganisms in stormwater should consistently be 99.9% (3.0 log) to 99.99% (4.0 log).

In this review, the efficiencies of structural treatment controls in removing microorganisms in stormwater were evaluated. In summary, BMP studies that were reviewed lacked consistent data reporting making comparisons between studies difficult, if not impossible. That said, removal efficiencies for structure treatment BMPs had removal efficiencies generally less than 90% (<1.0 log) for microorganisms in stormwater. Specifically, the studies reviewed indicate that retention ponds, detention basins, stormwater wetlands and sand filters had moderate removal efficiencies ranging from 0-99.0% (0 log to 2.0 log). Although no studies were found on the efficiencies of infiltration trenches and basins, infiltration may be a highly effective process for removing microorganisms in stormwater and therefore is worth future investigation. Additional data on the removal efficiencies of proprietary controls should also be considered as a research need.

Based on the results from the two stormwater advanced treatment case studies discussed, UV irradiation, in combination with pretreatment suspended sediment removal, should be considered as an effective treatment for the removal of indicator bacteria from dry-weather flows. That said, advanced treatment of stormwater should be approached cautiously. The construction of these facilities was completed due to the identification of high levels of microbial indicators, as opposed to pathogens, in dry weather flows from defined drainage areas. Limited evaluation of potential public health risk associated with these flows was conducted prior to construction of the facilities. These advanced treatment systems can also be expensive to construct and maintain (~$9 million for a 500,000 GPD system). Therefore, it is recommended that a cost-benefit analysis, which includes the consideration of risk, should be conducted prior to the construction of an advanced stormwater treatment facility.

Caution should be used when extrapolating these results of advanced stormwater treatment facilities in Southern California to wet-weather discharges (i.e., stormwater). Due to the high level of suspended solids typically present in stormwater, the efficiency of UV may be reduced due to particle occlusion. Additionally, the design capacity needed to treat stormwater discharges may be prohibitive, from both an engineering and cost perspective.

CHAPTER 6.0

A REVIEW OF MICROBIAL RISK CHARACTERIZATION METHODS AND A RECOMMENDED ASSESSMENT APPROACH FOR STORMWATER

1. Which waterborne pathogens pose the greatest risk to human health and what are the concentrations of these pathogens and indicator organisms observed in stormwater?
2. Are there defensible relations between indicator organisms and pathogens of public health concern in stormwater?
3. What is the fate of stormwater microorganisms (i.e., pathogens and indicator organisms) in the environment?
4. What microbial source tracking (MST) techniques are used to identify sources and assess potential public health risk?
5. What are the most effective stormwater microorganism control technologies and what are the estimated costs?
6. **What methods have been historically used to quantitatively characterize potential human health risk associated with exposure to waterborne pathogens and what is the most appropriate method for stormwater?**
7. What critical data are needed to enable the assessment of risk associated with exposure to pathogens in stormwater and what is the recommended approach?

6.1 Introduction

There are two general methods that are used to characterize the potential human health risks associated with exposure to pathogens in water; human health effects (epidemiologic) studies and risk assessments. Epidemiologic studies measure illness directly and associate the outcome measure with illnesses attributable to the exposure of interest. For example, a health effect study may look at a variety of outcomes such as gastrointestinal illness, skin rash, and ear infections that may be caused by a specific exposure to water. Risk assessments, on the other hand, generally estimate risk indirectly based on environmental measurements and numeric modeling.

Risk assessments are typically pathogen specific and therefore do not fully address all of the potential illnesses associated with a given exposure scenario that could include a number of pathogens. For example, researchers conducting a risk assessment may develop concentration estimates of pathogens or pathogen indicators in water. Those estimates could then be used as input to either a disease transmission model or a dose-response model to estimate the illnesses associated with a particular pathogen. In addition to having a characteristic uncertainty resulting from the limitation of pathogen-specific scenarios, risk estimates also depend on a number of model-based assumptions. These assumptions cause two potential sources of uncertainty: parametric uncertainty, due to limited knowledge of the values associated with the model parameters; and structural uncertainty, due to the model structure. For example, the fact that we only know the incubation period of an infection or the die-off rate of a given pathogen in the environment within a certain range might result in parametric uncertainty. To ascertain whether specific uncertainties are important with respect to regulatory or operational decision-making, researchers often employ sensitivity analysis. In this way, models become a valuable tool in directing both the decision-making process as well as future research needs.

There are also several important considerations associated with epidemiologic studies. First, well-designed studies can be extremely expensive. For example, an epidemiologic study may require millions of dollars to complete. Second, regulatory policy often must address scenarios, including different population groups (immuno-compromised groups, the elderly, etc.), different environmental settings (communities served by ground water vs. surface water), and different types of environmental loadings (point vs. non point sources) than those studied. Policymakers must apply the available study results to other scenarios, which sometimes can be quite difficult. Third, these studies may not be sufficiently sensitive to the risk levels that are relevant for regulatory activities. In an epidemiological study, the sample size determines the confidence of the risk estimates. Both cost and logistic difficulties in recruiting large numbers of participants limit the sample sizes in health effects studies.

Regulatory interests focused on risks associated with a particular contamination source, as opposed to a particular exposure scenario, add another layer of uncertainty. For example, the exposure pathways associated with microorganisms in stormwater are varied and not well defined. One can hypothesize a number of potential pathways, such as occupation-related exposures that result from direct contact with stormwater, or exposures where stormwater contributes to a waterbody used for recreation. Other pathways may also expose children to microorganisms in stormwater. Unfortunately, we know little about these exposure pathways and few health effects studies have focused on specifically on stormwater exposures.

The goal of this chapter is to provide a recommended approach to quantitatively assess a potential human health risk associated with exposure to stormwater. As background information, the chapter also provides overviews on health effect studies and microbial risk assessment methods.

6.2 Health Effects Studies

6.2.1 Introduction

The following sections provide an overview of different components of study designs for human health effects studies.

6.2.1.1 Outcomes Variables (Description of Health Effects)

Human health effects studies associated with waterborne pathogens generally focus on gastrointestinal illnesses (Wade et al., 2003). Researchers generally define these illnesses as some combination of diarrhea, vomiting, fever, nausea, and cramps. Highly credible gastrointestinal illness, for example, is defined as one of the following: vomiting; watery diarrhea; soft diarrhea and cramps occurring on the same day; or nausea and cramps occurring on the same day. Other health outcomes used in studies include eye infections; skin infections; ear, nose, and throat infections; and respiratory illnesses (Fleisher et al., 1998). A small portion of human health effects studies follow multiple outcomes.

6.2.1.2 Types of Study Designs

Environmental health effects studies can use a variety of study designs (Table 6-1). The following paragraphs provide brief descriptions of the most common designs.

The most rigorous type of human health effects study design is the intervention study. In recreational exposure trials conducted in the United Kingdom, Kay et al. (1994) randomly designated subjects as swimmers or nonswimmers. Investigators asked the swimmers to swim in a prescribed fashion. They measured water quality at or near the time of swimming. Traditionally, this has been considered the most rigorous study design because the investigator randomly assigns the swimmer status for each individual and therefore eliminates potential confounding factors that could bias the results (i.e., attributes of individuals, some of which may be casually related to both swimming and disease, are equally represented in both those that swim and those that do not swim). Because of the logistics involved, there may be more ethical concerns with this type of study than other type of studies.

Table 6-1. Types of Study Designs used to Determine Environmental Health Effects.

Type	Method	Advantages/ Disadvantages
Intervention	♦ Researcher assigns exposure ♦ Individuals followed over time post exposure to collect health outcome data	♦ Controlling exposure minimizes confounding ♦ Temporal follow-up establishes cause and effect ♦ Costly and time consuming
Prospective	♦ Researcher observes exposure ♦ Individuals followed over time post exposure to collect health outcome data	♦ Less expensive than an intervention trial ♦ Temporal follow-up establishes cause and effect ♦ Observing exposures introduces potential confounding
Cross Sectional	♦ Exposure and health data are collected at one point in time ♦ Data are collected at the individual level	♦ Less expensive and quicker to conduct than a prospective study ♦ Potential biases associated with confounding, cause and effect, and recall
Ecologic	♦ Exposure and health data are collected at the community level	♦ Inexpensive ♦ Can examine large population groups ♦ Potential biases associated with confounding

Prospective studies are often used in the recreational water studies (Pruss, 1998). In these studies, beach-goers were recruited and questioned about their swimming and exposure to water. They are contacted again three days to 1 month later and asked about health symptoms they experienced during this period. Water samples are collected periodically, usually at least once each interview day. Subjects are classified as swimmers and nonswimmers, and rates of illnesses in theses two groups are compared.

Other prospective studies follow athletes or focus on organized recreational events. Event participants comprise the exposed group. The unexposed group consists of bystanders, event organizers, or participants in a related event that does not involve swimming. The researchers contact the subjects after the event and ask them about the occurrence of illness. Water quality is also generally measures during the event.

There is a least one example of a cross sectional study where subjects were questioned about their recent illnesses at the same time as they were questioned about their swimming in the past four days (Wade et al., 2003). As with the prospective study, individuals are recruited at a recreational water site. Instead of following up with each person to see if they come down with an illness, the participants address exposure and illness that occurred in the past. The major issue with this design is recall bias.

In addition to study designs that focus on recreational water exposures to pathogens, a number of other study designs focus on the connection between health effects and exposure to water potentially contaminated with pathogens. For example, in a community-level study, (Naumova et al., 2003) employed time-series analyses to determine whether turbidity level in the drinking water effluent correlated with hospitalization rates and emergency room visits related to gastrointestinal illness. Similarly, Wade et al. (2004) examined the association of a flood event with gastrointestinal illness in the community. The difference in this study design compared with the ecologic studies was that the Wade team followed the cohort understudy more rigorously and defined the exposure event more precisely.

One ecological study examined the association between extreme precipitation and waterborne disease outbreaks by identifying monthly levels of precipitation for each reported outbreak (Curriero et al., 2001).

Various groups have extensively documented waterborne outbreaks. The CDC publishes a periodic description of identified outbreaks (Blackburn et al., 2004; Yoder et al., 2004). Though these outbreaks may not contain many cases, the reports do provide information about the more serious events associated with exposure to contaminated environments. These outbreaks are often viewed as a sample of those outbreaks that happen to be identified. It is generally assumed that most outbreaks go unnoticed.

Hepatitis A has been the focus of some cross sectional serological surveys of occupational exposures. In these studies blood samples are taken from a group of workers and risk factors are evaluated through questionnaires (Venczel et al., 2003; Weldon et al., 2000). One survey study was most closely related to the objectives of this report. The health status of postal service letter carriers was evaluated by reviewing how often the carriers took sick leave (Colford et al., 1999). Rainfall was used as a surrogate exposure measure to outflows from combined sewer and stormwater systems.

6.2.1.3 Weight of Evidence for Different Study Designs

Each of the study designs described above provides results that have different levels of weight of evidence. For example, a randomized trial is considered by many to be the best study design. Because the experimenter randomizes exposure of the participants, the results are less subject to confounding and therefore provide the strongest weight of evidence that a risk exists. The problem associated with this study design is that it is generally the most expensive epidemiological study to conduct and may be subject to additional ethical concerns. For these reason, few of these studies exist. In addition, both trials and prospective studies of environmental exposures are difficult to design, as it is often difficult to adequately control for exposure; i.e., it is difficult to insure that the unexposed group is truly unexposed and that the exposed group is exposed solely in the manner defined by the experiment. The randomized nature of the study design addresses this potential misclassification issue by randomly assigning exposure status, which means that any deviation from the defined exposure should occur equally in all groups (given the exposure assignment is independent of any factors associated with other sources of exposure), and therefore not confound the result. This type of exposure misclassification, however, will bias the risk estimate to the null and effectively decrease the statistical power of the study.

Prospective studies provide the temporal relationship between exposure and health effects that ensure the exposure occurs prior to an illness, however, the fact that the experimenter does not assign exposure status makes the results susceptible to confounding. For example, if those who happen to be swimming are healthier than those who choose not to swim, a real risk may be masked when comparing swimmers to non-swimmers. To get around this issue, information about potential confounding variables is generally collected, such as health status and age. These variables can be then used in the data analysis to provide adjusted estimates. In the area of recreational water studies, prospective studies are the most common study design used.

Cross-sectional studies provide weaker evidence of the causal relationship between an exposure and health risk. Since data are collected at one point in time, there is greater potential for a biased result compared with a study that collected data over time. There is no way of know the directionality of an association; i.e., did the act of boiling water result in individuals being protected from diarrheal disease or did the presence of diarrhea cause individuals to be more concerned about the water and therefore begin boiling the water. Cross sectional surveys, however, are inexpensive and easy to conduct. It is a good study design to explore possible risk factors that can be subsequently followed up by a prospective study or intervention trial.

Community level (or ecologic) studies are considered the weakest evidence towards establishing a causal relationship between an exposure and health risk. The chances for biases due to confounding or misclassification are high. As with the cross sectional surveys, they are conducted because they are inexpensive and relatively simple to conduct. As opposed to cross sectional studies, the data collected in an ecologic study is at the community level; e.g., the percent of people drinking boiled water and the percent of people with diarrhea. However, they are being used to infer risk at the individual level; e.g., is boiling water protective. Ecologic studies are a good hypothesis-generating tool and a good ecologic study can provide strong motivation for conducting a more comprehensive study.

6.2.2 Definition of Exposure

Exposures associated with stormwater are not well defined. Stormwater may affect the quality of recreational waters, may result in exposures to different occupations such as sewage workers and the general population through the contamination of drinking water or other undefined pathways. The following sections detail the different exposures used in health effects studies associated with some type of exposure to stormwater.

6.2.2.1 Types of Exposures

In recreational water studies, exposure has been divided into fresh and marine waters, because pathogens behave differently in these two types of water. Beyond this, categorization studies differed with respect to how exposure was assessed. Most studies determined swimming exposure through self-report or through proxy self-report (Stevenson, 1953; Cabelli et al., 1982; Cabelli, 1983b; Dufour, 1984; Ferley et al., 1989; Kueh et al., 1995;). A few studies reported having directly observed swimming behavior (Fleisher et al., 1993; Haile et al., 1999; Kay et al., 1994), while others determined exposure through participation in an event (Philipp et al., 1985; Fewtrell et al., 1992; Medema et al., 1995; Lee et al., 1997; Van Asperen et al., 1998). How swimming was defined also varied by study. Most required submersion of the head in the water, although some studies measured duration and intensity of exposure. The unexposed were often nonswimmers, however, some studies used swimmers in relatively higher quality water (as defined by water quality indicators).

Most other studies used surrogate markers to define exposure status. Occupational studies, for example, defined exposure by work activity (Khuder et al., 1998; Weldon et al., 2000; Venczel et al., 2003), whereas, drinking water studies defined exposure by whether the individual drinks tap water (Payment et al., 1991; Payment et al., 1997; Colford et al., 2005). Other studies defined exposure by place of residence (Morris et al., 1996; Morris et al., 1998).

6.2.2.2 Water Quality Measures

As pathogens are expensive to measure in the environment, most studies that were reviewed by the project team used surrogate microbial indicators in the exposure assessment. Most of the survey studies did not use water quality data in the analysis. Instead, exposure was determined indirectly.

The surrogate indicators can be divided into bacterial indicators of fecal contamination and nonfecal indicators. The fecal indicators included E coli, *Enterococcus*, fecal streptococci, fecal coliform, and total coliform. The nonfecal bacterial indicators included *Staphylococcus* spp., *Pseudomonas* spp., and Aeromonas spp. In addition, indicators of viral contamination included enteroviruses and bacteriophage.

6.2.3 Literature Review

Many health effects studies have examined the link between exposure to water and illness. These studies include those that focus on recreational water, drinking water and occupational exposures. Although these studies focus more on the human exposure component, as opposed to the source of contamination, they provide important information because stormwater can potentially be the source. Therefore, all of these health effects studies were thoroughly reviewed.

The literature search for this review was partially completed in a previous meta-analysis study (Wade et al., 2003). Additional publications were retrieved using several computerized databases: MEDLINE, BIOSIS, OLDMEDLINE, and EMBASE. The search terms included the key words "recreational water and health," "stormwaters, health effects," and "wastewater, health effects." Subject heading searches included "environmental pollutants, adverse effects" or "water pollution, adverse effects".

The project team organized the results of the literature review into two categories: recreational water studies (marine and freshwater), and ecologic, occupational, and outbreak studies. Although few published studies explicitly address stormwater exposure, their relevance lies in the plausible link between the health effects examined in the studies and exposures linked to microorganisms in stormwater.

6.2.3.1 Recreational Water Studies

This section provides a review of published recreational water studies that were recently evaluated by Pruss et al. (1998). All of these studies report risk in terms of "relative risk". Relative risk was generally defined as the ratio of the proportion ill in the exposed group divided by the proportion ill in the unexposed group. A relative risk of 1.5, therefore, suggests that those exposed were 1.5 times more likely to be sick than those who were unexposed. If everyone were exposed to some level of contamination, then a baseline exposure group would be chosen as the denominator. This baseline group is generally the group exposed to the lowest concentrations.

Seventeen studies were identified that focused on recreational swimming in marine waters. Thirteen of these studies were traditional prospective studies (Cabelli et al., 1982; Cabelli, 1983b; Fattal et al., 1986; Cheung et al., 1990; Alexander et al., 1992; Von Schirnding et al., 1992; Corbett et al., 1993; Pike, 1994; Kueh et al., 1995; Marino et al., 1995; Mcbride et al., 1998; Haile et al., 1999; Prieto et al., 2001;). One was a prospective event (Philipp et al., 1985), one was a cross sectional study (Foulon et al., 1983), and two were randomized trials (Fleisher et al., 1993; Kay et al., 1994). Five of these studies were conducted in the early to mid 1980s, nine were conducted in the early to mid 1990s, two were conducted in the late 1990s and one was conducted in 2001.

The bacterial indicators of fecal contamination considered in these studies were *E. coli*, fecal coliform and total coliform (Cabelli et al., 1982; Cabelli, 1983b; Cheung et al., 1990; Pike, 1994; Kueh et al., 1995). The strongest trend of increasing relative risk with increasing concentration of indicator was associated with *Enterococcus* (which was the only significant trend based on a linear regression). Estimates ranged from a relative risk of 1.58 with an exposure of 1-20 CFU/100 mL to 1.61 with an exposure of 21-54 CFU/100mL to 2.45 with an exposure of 69-104 CFU/100mL. Higher exposures (123-7460 CFU/100mL) did not show any additional increase in risk. The risk was statistically significant for all doses. For fecal coliform risks were significant for the lowest dose (RR = 1.4 with an exposure of 2-65 CFU/100mL) and for the highest dose (RR = 2.02 with an exposure of 1.46-2.77 CFU/100mL). There was no trend for the intermediate doses. For *E. coli*, risks were significant for all doses though there was no significant trend. The relative risk for the low dose range (2-54 CFU/100mL) was 1.8 and for the high dose range (320-5207 CFU/100mL) was 2.46. Relative risks based on total coliform were not significant.

The three nonfecal indicators used in these studies were *Staphylococcus* spp., *Pseudomonas* spp., and Aeromonas spp. (Cabelli, 1983b; Cheung et al., 1990; Kueh et al., 1995; Marino et al., 1995). Of these, only Aeromonas spp. resulted in significant risks of 3.86 and 3.27 for doses of 17 and 662 respectively.

The two indicators of viral contamination studied in marine waters were enterovirus and bacteriophage (Alexander et al., 1992; Von Schirnding et al., 1992; Pike, 1994; Haile et al., 1999). Combining studies, Wade et al. (2003) estimated a relative risk of 2.15 for an enterovirus exposure range of 0.53 – 4.7 PFU/10L. Only one study provided a detail analysis of bacteriophage and they found no association.

Ten studies conducted in freshwaters were identified. Of these 10 studies, six were traditional prospective studies (Stevenson, 1953; Dufour, 1984; Seyfried et al., 1985; Ferley et al., 1989; Lightfoot, 1989; Calderon et al., 1991), and four were prospective studies based on a sporting event (Fewtrell et al., 1992; Medema et al., 1995; Van Asperen et al., 1998; Lee et al., 2002). One study was conducted in the 1950's, four in the 1980's and four in the 1990's.

E. coli was the only indicator clearly associated with an increase in the relative risk of illness for freshwater exposure. The relative risk increased from 1.22 to 1.78 for exposure range increases from 45-170 to 185-204 CFU/100mL (Dufour, 1984; Lightfoot, 1989). Fecal coliform and *Enterococcus*, however, had significant risks for all dose ranges (Dufour, 1984; Ferley et al., 1989; Lightfoot, 1989; Seyfried et al., 1985). The relative risks ranged from 1.65 to 2.01. The nonfecal bacterial indicators showed no trends. *Staphylococcus* had significant relative risks of 4.3 and 2.73 for dose ranges of 14 and 36-45 CFU/100mL (Lightfoot, 1989), whereas *Pseudomonas* showed no significant risks (Ferley et al., 1989; Lightfoot, 1989) and Aeromonas showed a significant risk (2.4) at the high dose range of 1549 CFU/100 mL (Ferley et al., 1989).

The viral indicators, bacteriophage and enterovirus, both showed significant relative risks at all dose levels. For enteroviruses, the relative risks were 1.97 for a dose range of 0-0.4 PFU/100 mL and 4.11 for a dose range of 1 – 198.4 PFU/100 mL. Bacteriophage had risks of 2.42 and 2.8 for dose ranges of 0.07-20 and 1885 PFU/100 mL, respectively (Lee et al., 1997).

The meta-analysis conducted by Wade et al. (2003) provided a quantitative summary of the many studies conducted in the past 50 years. The studies taken as a group provide a much stronger weight of evidence towards the risks associated with recreational water swimming than any one study taken alone. The pooled analysis provides the best data to date to evaluate the existing regulations. The analysis suggests that the *Enterococcus* rule for marine waters and the *E. coli* rule for fresh waters are protective. Specifically, the risk associated with results from exposures above the U.S. EPA limits are significantly higher than the risks associated with results from exposures below the U.S. EPA limits. Clearly, this summary result may or may not pertain to a specific recreational site. More work is need to better characterize recreational sites beyond marine vs. freshwater.

6.2.3.2 Ecologic, Occupational, and Outbreak Studies

In addition to the recreational water studies reviewed in the previous section, a number of other health effects studies, such as ecologic studies of water quality and health, occupational studies and outbreak studies may also be relevant to stormwater. An ecologic study is one that does not collect data at the individual-level; i.e., there is no information on an individual's exposure status and illness. Rather data are collected at the community level (i.e., the number or percentage of individuals that are sick and the level of contamination within the community). Ecologic studies are useful in generating hypotheses for potential causes of disease; however, they are susceptible to bias and must be interpreted with caution. Occupational studies are of interest because they observe workers who are generally exposed to contamination at higher levels then the general population. Therefore, there is a greater chance of observing health effects. Finally, outbreak studies are of interest because illnesses are already identified. The outbreak investigation is therefore focused on the exposure that caused the cluster of illnesses.

A number of time series analyses have been conducted that have related daily fluctuations in drinking water turbidity with either hospital admissions or emergency visits for gastrointestinal illness. Two of these studies focused on the 1993 Milwaukee *Cryptosporidium* outbreak (Morris et al., 1998; Naumova et al., 2003). Morris et al. (1998) found that during the outbreak period, the correlation between drinking water turbidity and physician diagnosis was strongest at a lag of seven days in children and 8-10 days in adults. They hypothesized that these time lags were typical of incubation periods for *Cryptosporidium*. They also estimated stronger correlations before the outbreak period at a lag of eight days among children and nine days among adults. They hypothesized that waterborne cryptosporidiosis was occurring well before the documented outbreak. In a follow-up study the same study design was used but with a focus on the elderly. In this analysis they found the strongest correlations at temporal lags of five to six days, and a pronounced second wave that peaks at 13 days (Naumova et al., 2003).

In a similar set of analysis, Schwartz et al. (2000, 1997) analyzed data from Philadelphia. These data were not associated with a known outbreak. The first analysis focused on emergency visits and hospital admissions of children. For children age three and older, they found that an interquartile range increase in turbidity levels was associated with a 9.9% increase in gastrointestinal emergency visits four days later. For children two years old and younger the strongest association (5.9% increase) was found for a 10 day lag. For hospital admissions a similar pattern was found. An increase of 31% in children older than two years with a lag of 5-6 days, and an increase of 13% in younger children 13 days later.

The implications of these studies are not completely clear. There have been a number of critiques of this study design centered on both statistical and biological issues. The analysis consists of examining many different models consisting of different lag periods and different outcomes. The reasoning behind why certain models provided stronger correlation estimates occurred post analysis. In general, when conducting multiple model testing, there is the expectation that certain models would, by random selection, have correlation estimates that are stronger than the average estimates. The more models being tested the more likely one model will be associated with an outlier estimate even if there is no causal linkage. For this reason, these types of analysis are viewed as exploratory.

One potential concern with these studies is that often the results showed a strong correlation for a specific lag period, and weak correlation for a lag period the day prior or day after. These lag periods are interpreted as the average incubation period. Given the variable nature of incubation periods across individuals, a drastic change in correlation from one day to the next seems biologically improbable. These limitations further suggest that results should be interpreted with caution, particularly with respect to their relevance to stormwater.

One study analyzed the relationship between precipitation and waterborne disease (Curriero et al., 2001). Waterborne outbreaks were identified between 1948 and 1994. This particular analysis focused on situations in which two or more cases were associated with a drinking water source. Recreational outbreaks and outbreaks associated with sewage related issues were excluded from the analysis, since the focus of this study was on source waters and watershed contamination rather than accidental fecal releases and infrastructure problems in the distribution system. Each of the 548 outbreaks identified between 1948 and 1994 was geographically located at the centroid of one of the 2105 watersheds throughout the U.S. and was categorized as an outbreak caused by either surface or ground water contamination. Data were also collected on extreme precipitation events, defined as monthly precipitation for a given weather station that was in the highest 10% for the period between 1948 and 1994.

The analysis suggested that outbreaks were significantly associated with extreme precipitation events. Surface water events were most strongly correlated with precipitation events that occurred at the same time as the outbreak (0 month lag), whereas groundwater events were more strongly associated with precipitation events that occurred two months before the outbreak (2-month lag). The extent to which sewer overflow events impacted these results is not known.

Though this analysis may provide interesting hypotheses for further exploration, there is potential for bias and therefore does not have a strong weight of evidence. There are, however, numerous reported outbreaks that followed flooding that led to contamination of drinking water sources (Hunter, 2003). One study examined whether gastrointestinal illness was elevated during a flood period (Wade et al., 2004). A cohort was observed over time and provided weekly diaries of self-reported symptoms. During this follow-up period, severe flooding occurred in a community along the Mississippi River. Rates of illness were 1.29 times higher during the flood period compared with other periods. These effects were pronounced in children and with individuals that had direct contact with the floodwater.

Most occupational health effects studies follow a cross sectional study design. No occupational studies were found investigating exposure to stormwater. However, there were a number of occupational studies in the 1980s that compared sewage workers with controls (Cooper, 1991). Three U.S. EPA reports showed no difference in seroprevalence (Clark, 1980, 1981, 1985) between sewage workers and controls. Other studies, however, did find a difference in viral seroprevalence. Data were reported by Linneman (1984) suggesting a higher seroprevalence in coxsackie B5 antibodies, by Iftimovicic et al. (1980) suggesting higher adenovirus and parainfluenza antibodies, and by Skinhoj (1981) suggesting higher hepatitis A antibodies. Most of these studies stated that there was no illness observed during the survey or pathogens isolated in the feces. This, however, is not surprising given the limited number of participants and the cross sectional nature of the study. A few studies by Rylander presented data suggesting higher symptoms (fever, eye irritation, fatigue) in sewage workers compared to controls (Rylander, 1976; Mattsby and Rylander, 1978). Clark (1977) confirmed these findings and termed them the "sewage worker syndrome." He hypothesized that endotoxins caused this syndrome. One study detected *Giardia* in 16.5% of the sewage workers compared to 2.5% of the control group (Knobloch et al., 1983).

A few additional occupation exposure studies, examining the health risks of sewage workers, were conducted in the late 1990s and early 2000s. A study in Georgia examined the prevalence of hepatitis A, and suggested there was little or no additional risk among these workers (Venczel et al., 2003). Increased risks were found in a study in Texas between 1996 and 1997 (Weldon et al., 2000). They found that wastewater treatment plant workers had a higher risk of being infected with hepatitis A, compared with drinking water plant workers. These results were consistent with another study in Ohio in 1998 (Khuder et al., 1998). Colford et al (1999) studied postal workers and examined the association between rainfall and sick leave. The theory was that rainfall was a good surrogate indicator for combined sewer and stormwater overflows. No association was found.

The CDC periodically summarizes waterborne outbreaks every few years in Morbidity and Mortality Weekly Report (MMWR), which reviews two years of identified outbreaks. For example, in November 2002 a report summarized outbreaks identified in 1999 and 2000 (Lee et al., 2002). During that period, 39 outbreaks were associated with drinking water. These outbreaks caused 2,068 illnesses and two deaths. The etiology was identified in 22 (56%) of the 39 outbreaks. Twenty-eight were linked to groundwater sources, and of these 18 were associated with private or noncommunity wells. There were 59 outbreaks attributed to recreational water exposure affecting 2,093 persons, and the etiological agent was identified in 30 of these outbreaks.

There was an 11.8% increase in the reported drinking water outbreaks from 1999-2000 compared with 1997-1998. Groundwater outbreaks increased 87% comparing the same time periods, while recreational water outbreaks doubled. The recreational water outbreaks were most often associated with *Cryptosporidium* in treated water and by *E. coli* 0157:H7 in freshwater. Similar results were reported for the time period 2001 to 2002 for recreational water outbreaks (Blackburn et al., 2004). In a separate summary of recreational water outbreaks from the time period of 1971-2000, it is reported that most were attributed to bacteria or protozoa. *Shigella* and *E. coli* 0157H7 were often associated with freshwater outbreaks, whereas *Cryptosporidium* and *Giardia* were primarily associated with treated water sources. Important causes of contamination were identified as other bathers, sewage discharge and wild or domestic animals.

The source of the contamination for the outbreaks described above is seldom identified and therefore there are little data associating outbreaks specifically with stormwater. When a more distal cause is identified, however, it is often related to sewage and occasionally to sewerage-contaminated stormwaters. For example, in 1998, a drinking water outbreak of cryptosporidiosis that occurred in Brushy Creek, Texas was linked to storms that led to sewage contamination of wells and creeks (Barwick et al., 2000). In 1988-1989, a hepatitis A epidemic in Ococee, Florida was associated with sewage overflows (Vonstille et al., 1993). Of the 18 index cases, each had a history of contact with sewage-contaminated stormwater and no other known contact with the infection.

6.3 Microbial Risk Assessment

6.3.1 Introduction

Risk analysis consists of three principal components: risk assessment, risk management, and risk communication. Within this construct, risk assessment characterizes and estimates, either qualitatively or quantitatively, the potential adverse health effects associated with exposure of individuals or populations to hazardous materials and situations (Hoppin, 1993; National Research Council, 1983). The risk management component weighs policy alternatives in light of the results of risk assessment and, if required, selects and implements appropriate control options, including regulatory measures. The risk communication component is the exchange of information and opinions concerning risk and risk management among risk assessors, risk managers, consumers and other interested parties (World Health Organization, 1999). This section focuses on the risk assessment component of the risk analysis process.

Microbial risk assessment (MRA) evaluates the likelihood of adverse human health effects that can follow exposure to pathogenic microorganisms or to a medium in which pathogens occur (ILSI, 2000). To the extent possible, the MRA process includes evaluation and consideration of quantitative information; however, it also employs qualitative information as appropriate (World Health Organization, 1999).

Quantitative risk assessment has been used since the 1970s to assess human health effects associated with exposure to chemicals (Carter et al., 1983). In 1983, the National Research Council (NRC) formalized the principles, processes and methods for carrying out risk assessments for chemical agents (National Research Council, 1983). The NRC identified four steps for chemical risk assessments: hazard identification, dose-response assessment, exposure assessment, and risk characterization. Many of the early MRAs employed the NRC conceptual framework in conducting assessments (Haas, 1983; Regli et al., 1991; Rose et al., 1991).

As the field of MRA developed, complexities associated with modeling infectious diseases unique to pathogens emerged, such as immunity and person-to-person transmission of infection. It became apparent that a chemical-style assessment might not be the only appropriate method for assessing risk of human infection following exposure to pathogens (ILSI, 2000). To address this concern, the U.S. EPA Office of Water sponsored a series of workshops to develop a conceptual framework to assess the risks of human infection associated with pathogenic microorganisms. Those workshops resulted in a published framework (ILSI, 2000) that was then tested in two case studies (Teunis and Havelaar, 1999; Schaub, 2004). Three principal components comprise the U.S. EPA framework for assessing the risk of human infection following exposure to water- and food-borne pathogens: problem formulation, analysis, and risk characterization. At this time, MRAs employ both the NRC and U.S. EPA frameworks. A brief summary of those frameworks follows.

6.3.2 NRC Risk Paradigm for Microbial Risk Assessment

This section summarizes the four-step NRC risk paradigm, as applied to microbial risk assessment.

6.3.2.1 Hazard Identification

For microbial agents, the purpose of hazard identification is to identify the microorganisms or the microbial toxins of concern. Relevant data sources for identifying hazards include scientific literature, databases, and expert opinion. Hazard identification often involves one or more of the following (World Health Organization, 1999):

- review of clinical studies;
- epidemiological studies and surveillance;
- laboratory animal studies;
- investigations of the characteristics of microorganisms and the interaction between microorganisms and their environment; and
- studies of analogous microorganisms and situations.

6.3.2.2 Dose-Response Assessment

Dose-response assessment provides a quantitative or qualitative description of the likelihood, severity, or duration of adverse effects that may result from exposure to a microorganism or its toxin. Different conditions, such as infection or illness, may have different dose-response relationships. In the absence of dose-response data for factors such as infectivity, researchers may use risk assessment tools such as expert elicitations to characterize the host's response to a dose of pathogens (World Health Organization, 1999).

Important factors relative to the microorganism include:

- the virulence and infectivity of microorganisms can change, depending on their interaction with the host and the environment;
- genetic material can pass between microorganisms, leading to the transfer of characteristics such as antibiotic resistance and virulence factors; and
- low doses of some microorganisms can in some cases cause a severe effect.

Important factors relative to the human host (World Health Organization, 1999):

- genetic factors;
- increased susceptibility due to breakdowns of physiological barriers;
- individual host susceptibility characteristics such as age, pregnancy, nutrition, overall health, medication status, concurrent infections, immune status and previous exposure history; and,
- population characteristics such as population immunity, and access to and use of medical care.

6.3.2.3 Exposure Assessment

An exposure assessment describes the magnitude or probability of actual or anticipated human exposure to pathogenic microorganisms or microbiological toxins. For microbiological agents, researchers might base exposure assessments on potential contamination in water by a particular agent or its toxins.

Other factors considered in performing exposure assessments include (World Health Organization, 1999):

- the frequency of human exposure to the pathogenic agents;
- the duration of exposure;
- the associated concentrations of those pathogens over time; and
- socio-economic status, ethnicity, seasonality, population demographics, regional differences, and consumer preferences and behavior.

Researchers have also evaluated the potential impact of environmental conditions and water treatment reliability.

6.3.2.4 Risk Characterization

Risk characterization represents the integration of the hazard identification, dose-response assessment, and exposure assessment components to obtain a risk estimate. The risk characterization process results in a qualitative or quantitative estimate of the likelihood and severity of the adverse effects that could occur in a given population. It includes a description of the uncertainties associated with that estimate.

Risk characterization depends on available data and expert interpretation of those data. The integration of quantitative and qualitative data may permit only a qualitative estimate of risk. The degree of confidence in the final estimation of risk will depend on the variability, uncertainty, and assumptions identified in all previous steps (World Health Organization, 1999). Ideally, risk management considerations may require differentiation of uncertainty and variability. However, experience indicates that for MRAs, it is difficult or impossible to consider them independently.

6.3.3 U.S. EPA Framework for Microbial Risk Assessment

The U.S. EPA Microbial Risk Assessment (MRA) framework (ILSI, 2000) is conceptually similar to the NRC paradigm for human health risk assessments (National Research Council, 1983) and the ecological risk assessment framework (U.S. EPA, 1992). The U.S. EPA framework emphasizes the iterative nature of the risk assessment process and allows wide latitude for planning and conducting risk assessments in diverse situations. This framework consists of three principal components: problem formulation, analysis, and risk characterization. The U.S. EPA subdivides the analysis phase into the characterization of exposure and human health effects.

The problem formulation stage involves all stakeholders and identifies the purpose of the risk assessment; the critical issues; and how the results might help protect public health. In this stage, researchers identify and describe the exposure and potential health effects, then develop a conceptual model. This model serves as a starting point for the analysis phase of the risk assessment. Later, it functions as an interactive tool, along with components developed in the analysis phase, to initiate risk characterization.

In the analysis stage, researchers summarize information about both exposure and health effects. This may include quantitative and qualitative data, expert opinion, and other information. This compilation results in exposure and host/pathogen profiles that explicitly identify the data integrated into the risk characterization and the associated assumptions and uncertainties. These two elements, while separate, must also interact to ensure that the results are compatible.

The final stage, risk characterization, results in a statement of the likelihood, types, and magnitude of effects that the exposed population may experience in the exposure scenario. The risk characterization statement includes all of the inherent assumptions and uncertainties. Often, the risk characterization phase integrates the data through parameterization of a mathematical model, numerical simulation, and interpretation.

6.3.4 Risk Characterization Methods

Quantitative methods to characterize human health risks associated with exposure to pathogenic microorganisms were first published in the 1980s (Cooper, 1986; Haas,1983). Over the last 20 years, the field of microbial risk assessment has been developing and maturing. Since that time, dose-response relations have been developed for various pathogenic microorganisms (McBride et al., 2002), and microbial risk assessment investigations have been carried out for a number of those pathogens in water, food and other media. This section reviews and summarizes the dose response and risk characterization methods that have been most commonly employed in MRAs for waterborne pathogens.

6.3.4.1 Evaluation of Dose-Response Relations

The dose-response assessment provides a qualitative or quantitative description of the likelihood of adverse effects that may result from exposure to a microorganism or its toxin. For MRA, the adverse health effects most commonly of interest are infection and illness. Infection may be defined as the invasion, colonization and multiplication of a pathogenic microorganism (Teunis et al., 1996). Since humans may come into contact with a wide range of pathogen concentrations, it is therefore useful to develop a functional relationship between a quantitative measure of microorganism ingestion and the risk of infection. A dose is defined as a quantitative measure for the intensity of exposure of the host to the pathogen of interest. The units of dose are usually given as the number of organisms ingested. By studying the effects of various doses, it is possible to determine a dose-response relationship between the dose and the frequency of infection within the exposed population of hosts. Researchers have studied quantitative descriptions of the dose-response relationships of organisms to provide insight into the risk of becoming infected after the ingestion of a certain dose of organisms (Teunis et al., 1996).

Dose-response data from the results of clinical trials have been reported in the literature (Mcbride et al., 2002) for *Campylobacter* (Black, 1988), adenovirus 4 (Couch, 1966; Couch, 1969), echovirus 12 (Akin, 1981), *Salmonella* (McCullough and Eisele, 1951), *Cryptosporidium* (Dupont, 1995; Chappell, 1996; Moss, 1998; Okhuysen, 1998; Okhuysen, 1999; Messner, 2001), *Giardia lamblia* (Rendtorff, 1978), Vibrio cholera (Hornick, 1971), rotavirus (Ward et al., 1986), poliovirus (Lepow, 1962; Katz, 1967; Minor, 1981; Regli, 1991), Escherichia coli O157:H7 (Powell, 2000), and coxsackieviruses B4 and A21 (Couch, 1965; Suptel, 1963). Generally, these data were obtained from feeding studies of healthy adults. Thus, certain portions of population including children, the elderly, and individuals with compromised immune systems are not well represented by these data.

Typically, researchers fit the reported dose-response data to models that relate the probability of infection to the mean dose ingested. Some studies have also investigated illness as an end-point. However, the conditional modeling of illness, given infection, has proven difficult (Teunis and Havelaar 1999). The most common, although not the only, models used to relate an ingested dose to infection are the exponential and beta-Poisson models (Haas et al., 1999).

The exponential model is based on the following assumptions (Haas et al., 1999):

- microorganisms are distributed in water randomly and thus, follow the Poisson distribution;
- for infection to occur, at least one pathogen must survive within the host; and
- the probability of infection per ingested or inhaled organism is constant.

Mathematically, the probability of infection (Pinf) as expressed by the exponential model is as follows:

$$P_{inf} = 1 - e^{-rN}$$

In the exponential model, each microorganism has the same fixed probability (r) of surviving and reaching a host site at which infection may result. Under this model, the dose required to cause infection in half the exposed population is N50 = –ln(0.5)/r. The dose-response relation for many protozoans and viruses tend to follow this model. The biological implication of this model is that differential susceptibility in the challenged population tends not to be strong (McBride et al., 2002).

The beta-Poisson model depends on similar assumptions, except that the third assumption (that the probability of infection per ingested organism is constant) is relaxed. This model allows the probability of infection per ingested or inhaled organism to vary with the population. In this model, the probability of surviving and reaching a host site ("r" in the exponential model) is beta distributed. Thus, the model contains the two parameters (α and β) of the beta distribution. The most commonly used approximation to the beta-Poisson model is:

$$P_{\inf} = 1 - \left(1 + \frac{N}{\beta}\right)^{-\alpha}$$

Unfortunately, in this approximation to the beta-Poisson model, α does not have an obvious physical interpretation. It is a shape parameter governing the steepness of the dose-response curve; the larger its value, the steeper the curve (McBride et al., 2002). The derivation of the approximation to the beta-Poisson model requires that $\beta >> \alpha$ and become poorer at small values of β or large values of N. In practice, this condition is not always met. The beta-Poisson is linear at low doses and is always shallower than the exponential model. However, as $\alpha \rightarrow \infty$, the beta-Poisson model approaches the exponential model (Haas et al. 1999).

The beta-Poisson model describes many bacteria and some viruses. For organisms whose dose-response relations this model describes, the biological implication is that there is substantial differential susceptibility in the challenged population (McBride et al., 2002). For a compendium of critically analyzed dose response curves, refer to Haas et al. (1999) or McBride et al. (2002).

New methods for dose-response assessment relying on Bayesian approaches have appeared in the literature over the last several years (Englehardt, 2004; Messner, 2001). These methods attempt to address a major limitation of likelihood-based approaches; specifically, that data validation is not possible for doses of pathogens that are consistent with public health goals. A detailed description of these methods is beyond the scope of this report.

6.3.4.2 Risk Assessment Methods

Soller et al. (2004b) recently conducted a literature review to document the status, advantages, and limitations of different types of microbial risk assessment techniques. The literature review of approximately 1,100 citations indicated that at the broadest level there was a distinction between direct estimates of risk or illness using epidemiological data and indirect estimates using models. Direct estimates entail collecting infection or disease outcome data; for example, prospective studies or outbreak investigations. Indirect estimates employ exposure data, as input to numerical models, to compute estimates of illnesses.

Based on the available literature, it appears that direct methods are most commonly used to assess the public health impact associated with a specific and known (or identifiable) exposure pathway. These methods may not provide the regulatory and management information needed to make decisions regarding changes in environmental conditions. For this purpose, indirect methods can play a useful role.

The literature review indicated that MRA methodologies vary primarily in the manner in which they address the unique properties of an infectious disease transmission system. The fundamental difference between these risk assessment techniques is that NRC paradigm models (static models) do not account for the unique properties of a dynamic infectious disease process (Table 6-2). In static models, the number of individuals assumed susceptible to infection is not time varying, whereas in dynamic models that number is time varying.

Assessments using a static model for evaluating microbial risk typically focus on estimating the probability of infection or disease to an individual because of a single exposure event. These assessments generally assume that multiple or recurring exposures constitute independent events with identical distributions of contamination (Regli et al., 1991). Secondary transmission and immunity are typically not considered, assumed to be negligible, or effectively cancel each other out.

Table 6-2. Comparison of Static and Dynamic Risk Assessment Models.

Static risk assessment model	Dynamic risk assessment model
The # of individuals susceptible to infection is time invariant	The # of individuals susceptible to infection is time variant
Direct exposure (environment-to-person)	Direct (environment-to-person) and indirect exposure (person-to-person)
Individual-based risk	Population-based risk
Potential for secondary transmission of infection or disease is typically not considered.	Potential for secondary or person-to-person transmission of infection or disease is considered.
Immunity to infection from microbial agents is typically not considered.	Exposed individuals may not be susceptible to infection or disease because they may already be infected or may be immune from infection due to prior exposure.
Dose-response function is the critical health component.	The dose-response function is important; however, factors specific to the transmission of infectious diseases may also be important.

Static MRA models assume that the population may be categorized into two or three epidemiological states: a susceptible state, an infected, and sometimes a diseased state. Susceptible individuals are exposed to the pathogen of interest. The probability of the exposed individuals moving into the infected/diseased state is governed by the dose of the pathogen to which they are exposed and the infectivity of the pathogen (Figure 6-1). In Figure 6-1, solid lines represent the movement of individuals from one epidemiological state to another, and dotted lines represent the movement of pathogens. Although a number of environmental sources may expose humans to pathogens, static models typically assume that susceptible individuals are exposed to pathogens from the specific pathway under consideration for the investigation. They therefore do not include the potential interactions and implications of multiple routes of exposure.

The probability that a susceptible individual will become infected or diseased is a function of the dose of pathogens to which that individual is exposed. When individuals are exposed to pathogens from an environmental source, they move with a given probability to an infected or diseased state. This probability dose-response function is labeled $P_{(inf)}$ in Figure 6-1. The dose is typically calculated by estimating two quantities: the concentration of pathogens at the exposure site and the volume of water ingested. This dose quantity is then input into the dose-response function, allowing one to estimate the probability that an exposed individual will become infected or diseased.

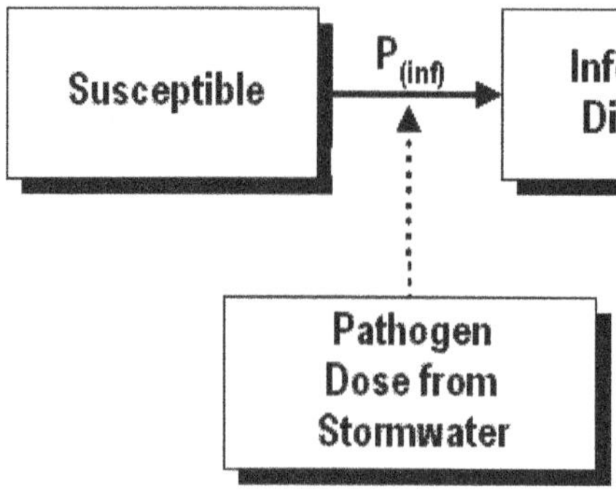

Figure 6-1. Conceptual Model for a Static Risk Assessment.

The pathogen-specific dose-response function that represents the probability of infection $P_{(inf)}$ contains a summary of the critical health effects information required for the static model. The probability of infection following exposure to a virulent pathogen depends on several host and pathogen-specific factors. One can view the interaction between a pathogen and the host as a series of conditional events, each of which must occur to result in infection (Teunis et al., 1996). The infection status depends on a number of factors such as:

- the number of organisms that enter the host;
- the host's ability to inactivate these organisms;
- the number of organisms that can withstand the host's local immune defenses, adhere to mucosal surfaces, and multiply in order to infect the host; and
- the variation in pathogen virulence and host susceptibility.

The probability of infection multiplied by the number of exposed individuals gives the expected number of infected individuals for the exposure scenario under consideration.

A dynamic risk assessment model makes assumptions that divide the population into epidemiological states. Individuals move from one state to another based on epidemiologically relevant data (duration of infection, duration of immunity, etc). Only a portion of the population is in a susceptible state at any point in time, and only those individuals can become infected or diseased through exposure to microorganisms. The probability that a susceptible person moves into an exposed state is governed by the dose of pathogen to which they are exposed; the infectivity of that pathogen; and the number of infected/diseased individuals with whom they may come into contact. For both dynamic and static representations of the disease process, infectivity as a function of dose (estimated using a dose-response function) is an important factor in estimating risk. The dose-response function is also important in a dynamic microbial risk assessment model, however, other important factors may include person-to-person transmission, immunity, asymptomatic infection, and incubation period.

Accounting for these additional factors when estimating risks associated with exposure to pathogenic microorganisms requires a more sophisticated mathematical model than the static model shown conceptually in Figure 6-1. When using a dynamic disease transmission model, one can account for attributes specific to the transmission of infectious diseases. Depending on the infectious disease processes that are important, the dynamic model may include more or fewer components, and therefore vary in complexity. For example, a dynamic model may account for person-to-person transmission, immunity, incubation, and asymptomatic infection, as illustrated in Figure 6-2 (Soller et al., 2004b). The solid lines in Figure 6-2 represent the movement of individuals from one epidemiological state to another, and the dotted lines represent the movement of pathogens.

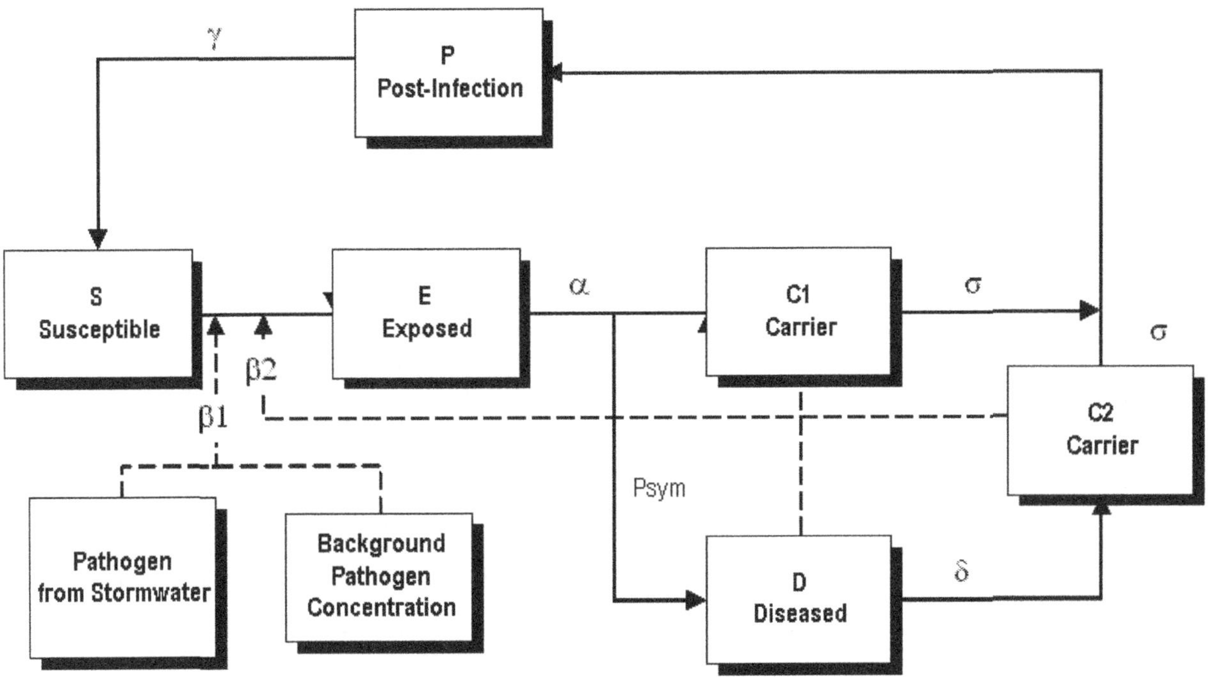

Figure 6-2. Conceptual Model for a Dynamic Risk Assessment Addressing Exposure to Pathogens from Stormwater

Figure 6-2 shows the population separated into six epidemiological states. Table 6-3 provides a summary of the epidemiological states employed in this dynamic model. Table 6-4 summarizes rate parameters specifying the movement between epidemiological states, showing the parameters as Greek letters.

Table 6-3. Epidemiological States for a Representative Dynamic Model.

Figure 3 Label	Epidemiological States	Description
S	Susceptible	Individuals not infected and not protected from infection.
E	Exposed	Individuals exposed to a pathogen, but not yet infected.
C1	Carrier 1	Individuals infected but do not have symptoms of disease.
D	Diseased	Individuals infected and have symptoms of disease.
C2	Carrier 2	Individuals diseased, no longer exhibit symptoms of disease, but still infected.
P	Post-Infected	Individuals neither infected nor symptomatic, and have resistance to infection.

Table 6-4. Rate Parameters for a Representative Dynamic Model.

Symbol	Description
α	Rate of movement from an exposed state to a carrier (infectious and asymptomatic) state or a diseased state (infectious and symptomatic). 1/□ corresponds to the latency period before infection for the pathogen of interest.
σ	Rate of movement from a carrier state to a post infection state,¯ı□ corresponds to the duration of infectiousness, or equivalently, the duration of asymptomatic shedding of pathogen in feces.
δ	Rate of movement from a diseased state (infectious and symptomatic) to an asymptomatic (carrier) state. $1/\delta$ corresponds to the duration of symptoms during infection.
γ	Rate of movement from a post-infection state (not infectious, asymptomatic, and not susceptible to infection) to a susceptible state. 1/□corresponds to the duration of immunity or protection from infection.
$\beta 1$	Rate of movement from a susceptible state to an exposed state due to exposure to pathogens from an environmental source (i.e. not person-to person transmission). Function of the number of pathogens to which an individual is exposed and the infectivity of the pathogen of interest. A dose-response function that is comprised of one or two dose-response parameters describes the infectivity quantitatively.
$\beta 2$	Rate of movement from a susceptible state to an exposed state due to exposure to pathogens from secondary (person-to-person or person-to-environment-person) transmission.
P_{sym}	Probability of a symptomatic response. Clinical data describing the proportion of infected individuals that develop symptoms.

The model shown in Figure 6-2 is called a dynamic model because the number of people in each epidemiological state varies over time. The dynamic model is a more comprehensive mathematical rendering of the static model. Under a specific set of assumptions the two models are equivalent. A comparison of Figures 6-1 and 6-2 indicates that the two models would be equivalent when:

- the background concentration of the pathogen (or, equivalently, the endemic level of infection/disease) in the population is zero;
- the duration of infection and disease approaches zero; and
- infection and disease do not confer immunity, or the duration of immunity approaches zero.

In categorizing the epidemiological status of the population, individuals are considered infected if they are shedding pathogens in their feces or have an immunological response. People are considered diseased if they exhibit any of the clinical symptoms related to the specific pathogen of interest; for example, diarrhea or vomiting.

Dynamic MRA models can take two main forms: deterministic or stochastic. In the deterministic form, the model is expressed as a set of differential equations that have defined parameters and starting conditions which determine the rate of transfer of individuals from one epidemiologic state to another. This type of model is most suitable for large populations of individuals randomly interacting with one another (Eisenberg et al., 1998; Soller et al., 2003). In the stochastic form, the model incorporates probabilities at an individual level and is evaluated by an iterative process such as Markov Chain Monte Carlo analysis. Stochastic model forms are best suited to small populations with heterogeneous mixing patterns.

A variety of MRA model forms can be used to characterize infectious disease transmission and evaluate the potential for effective interventions. Particular characteristics of each model form capture different aspects of the disease transmission system. However, it is unrealistic to presume that one model form is most appropriate for all waterborne microbial risk assessments. (Soller et al., 2004b) demonstrated that for exposures to microbes from reclaimed water applications, the selection of an appropriate model form (static or dynamic) could be made based on as few as three to four parameters. That investigation also clearly demonstrates that no model form will be appropriate for all possible combinations of exposures and pathogens of potential interest.

The selection of a model type does involve tradeoffs. Biological or demographic "realism" comes at the cost of analytical complexity that distances the model from available data. Further, each model form involves certain types of assumptions that may or may not be realistic or appropriate for a particular situation. Recognizing that different applications may require different model forms and analytical approaches, Koopmans et al. (2001) suggest an analysis strategy involving a hierarchy of simple to increasingly complex models. These could be traversed to make MRA analyses more realistic while remaining mathematically tractable. It seems quite reasonable to anticipate that the issue of model complexity for microbial risk assessments will receive substantial attention.

6.4 Evaluating Risk Associated with Exposure to Pathogens in Stormwater

6.4.1 Pathogens of Public Health Concern

Conducting an assessment of public health risk associated with exposure to pathogens requires the selection of a representative pathogen or pathogens (hazard identification). Although a wide range of pathogens have been identified in raw wastewater and to a lesser extent in stormwater, Mead et al. (1999) determined that relatively few pathogens cause the majority of the non-foodborne illnesses caused by known pathogens in the U.S. The pathogens that have been reported to be responsible for the vast majority of illnesses in the U.S. ("pathogens of public health concern") from all sources should be the focus of future work.

To identify the pathogens of public health concern, research conducted by the Centers for Disease Control (CDC) was employed by Mead et al. (1999). In characterizing food-related illness and death in the U.S., the authors estimated the annual total number of illnesses caused by known pathogens (adjusting for the fact that many illnesses are not reported) in the U.S. In summary, those estimates indicate an estimated 38.6 million cases of illness occur annually in the U.S., with 5.2 million cases from bacterial pathogens, 2.5 million from parasitic pathogens, and 30.9 million from viral pathogens (see Table 2-1).

Further inspection of Mead et al. (1999) estimates indicates that approximately 85-90% of all non-foodborne infections in the U.S. are thought to be caused by viral pathogens (i.e., enteric viruses). The relative importance of viral pathogens in waterborne transmission of disease is supported by data from the World Health Organization (1999) and research conducted over the last 20 years on exposure to waterborne pathogens through recreational activities (Cabelli, 1983b; Levine and Stephenson, 1990; Palmateer et al., 1991; Sobsey et al., 1995; Fankhauser et al., 1998;). Norwalk-like viruses account for 23 million illnesses each year, of which 60% are estimated to be non-foodborne. Rotavirus accounts for 3.9 million illnesses each year, of which 99% are non-foodborne (Mead et al., 1999).

Giardia lamblia and *Cryptosporidium parvum* appear to be the most important waterborne parasitic pathogens. Based on (Mead et al., 1999) estimates, *Giardia lamblia* causes two million illnesses per year, of which 90% are non-foodborne, and *Cryptosporidium parvum* causes 300,000 illnesses each year, of which 90% are non-foodborne.

Many of the important bacterial illnesses in the U.S. are primarily transmitted through food. Of those bacterial pathogens which account for the non-foodborne illnesses in the U.S., *Salmonella* (*Salmonella* spp. and non-typhoidal), *Shigella*, and *Campylobacter* spp. together account for approximately 4.3 million of the 5.2 million annual bacterial illness cases in the U.S, respectively (of which, approximately 95%, 20%, and 80% are foodborne, respectively) (Mead et al., 1999).

For the purposes of this investigation, the primary criteria used to select pathogens as being representative of stormwater includes: 1) the likely presence of the pathogen in stormwater based on the potential of fecal contamination, and 2) the estimated importance of the pathogen relative to the total number of illnesses in the U.S. and likely importance relative to the number of cases of waterborne disease in the U.S. From the list of pathogens identified by the U.S. EPA (2002a), and CDC's estimated illness levels in the U.S., the pathogens of public health concern selected for this investigation are: Norwalk-like viruses, rotavirus, and hepatitis A (Human Viruses), *Cryptosporidium parvum* and *Giardia lamblia* (Protozoa), and *Salmonella* spp., *Shigella* spp. *Campylobacter* and pathogenic *E. coli* (Bacteria).

Given that the above selected pathogens of public health concern are responsible for a large majority of waterborne infections in the U.S., it is assumed for this investigation that the potential public health risk associated with exposure to these pathogens provides a representative estimate of the overall human health risk through contact with stormwater.

6.4.2 Recommended Risk Characterization Method for Evaluating Pathogens in Stormwater

As reviewed at the beginning of the chapter, risk characterization is generally conducted using one of two methods; epidemiological studies of human health or MRA. The literature search conducted by the project team identified epidemiological studies that focused on health effects associated with non-point source microbial contamination. The literature review indicated that few epidemiological studies have specifically addressed this topic. Specifically, four investigations were identified; one study addressing risks in freshwater (Calderon et al., 1991) and three studies addressing risks in marine waters (Dwight, 2004; Colford, 2005; Haile, 1999). Calderon et al. (1991) concluded that the currently recommended bacterial indicators are "ineffective for predicting potential health effects associated with water contaminated by non-point sources of fecal pollution." The results from the study conducted by Colford et al. (2005) were consistent with that finding. The Santa Monica Bay study (Haile et al., 1999) indicated that the risks of several health outcomes were higher for people who swam at storm drain locations, compared to those who swam further from the drain. However, the list of health outcomes that were more statistically significant (fever, chills, ear discharge, cough and phlegm, and respiratory) did not include gastrointestinal illness, which is curious because the vast majority of epidemiological studies worldwide suggests a causal dose-related relationship between gastrointestinal symptoms and recreational water quality measured by bacterial indicator counts (Pruss, 1998). Dwight et al. (2004) found that surfers in an urban environment reported more symptoms than their rural counter-parts, however, water quality was not specifically evaluated in that study. Therefore, research has shown the difficulty in drawing conclusive inferences from these types of studies about specific impacts associated with exposures to pathogens in stormwater.

At this stage, the use of risk assessment to assess public health concerns associated with stormwater is a valuable alternative to conducting epidemiological studies. Previous WERF research by Soller et al. (2004b) enables us to evaluate the applicability of MRA techniques for estimating the public health risk from exposure to microorganisms via stormwater. The infectious disease process in a population is, fundamentally, a dynamic process. Therefore, a dynamic model provides the most rigorous and scientifically defensible approach for quantifying the infectious disease process. However, under certain conditions and assumptions, dynamic and static models may yield similar results. These conditions are of particular interest, because modeling the transmission of infectious diseases as a static process requires substantially less data and mathematical sophistication than modeling the process dynamically. In selecting a microbial risk assessment model, however, one must be careful to ensure that any simplifying assumptions are in fact appropriate from an epidemiological perspective.

Previous WERF research by Soller et al. (2004b) determined the conditions under which specific realizations of static and dynamic models predicted similar and substantially different estimations of risk. The subject of that investigation was the identification of risks from exposure to reclaimed water. Inferences are drawn here, to the extent feasible, for stormwater exposures based on similarities of exposures to pathogens.

The fundamental premises upon which the model evaluation results were based are as follows:

- Under the conditions in which the two models predict similar estimations of risk, the static model is more appropriate, as it is simpler yet yields similar results.
- Under the conditions in which the two models predict substantially different estimations of risk, the dynamic model is more appropriate, as one or more infectious disease processes impact the assessment enough to impact the assessment of risk.

To differentiate between the conditions under which the static and dynamic models predict similar and substantially different estimations of risk for the specific exposure scenario under consideration, a series of numerical simulations were set up to explore the range of feasible parameter combinations. The exposure represented a range of pathogenic microorganisms via reclaimed water applications. A sensitivity analysis was then conducted to determine what parameters in the model impacted the predicted difference in infection incidence between the two models.

The results of the research suggest that the parameters that most strongly impacted the difference in predicted incidence between the static and dynamic model investigations were, in order of decreasing importance:

- dose of pathogen;

- exposure intensity;
- dose-response parameter beta;
- dose-response parameter alpha ; and,
- duration of infection.

Critical combinations of the model parameters identified above are summarized graphically in Figure 6-3. In this graphical summary of the model evaluation, three 3x3 grids are presented, one for each of three threshold levels of tolerable incidence difference[13]:

- Threshold A. Difference in Predicted Incidence =< 10/100,000 per year
- Threshold B. Difference in Predicted Incidence =< 1/100,000 per year
- Threshold C. Difference in Predicted Incidence =< 0.01/100,000 per year

For each grid, the levels of exposure intensity vary by row. Exposure intensity is the proportion of the population exposed times the frequency of exposure: the bottom row is low exposure intensity (3x10-5), the middle row is medium (10-3), and the top row is high (0.1).

The dose levels vary by column in each grid. The dose represents the number of organisms ingested by an individual via the exposure under consideration. The left column represents a low dose (10-8), the middle column is medium dose (10-4), and the right column is high dose (1.0).

The levels of the dose-response parameter beta vary by shape. In each grid, low values for the dose-response parameter beta (0.21) are represented by a circle (typically viruses and some highly infectious bacteria), medium (0.39) by a square (typically parasites), and high (444,000) by a triangle (typically less infectious bacteria).

The conditions resulting in low difference in predicted incidence between the static and dynamic models (where low is defined for each grid) are represented in gray, and those not resulting in low difference in predicted incidence are represented in white.

For example, assume that we are interested in an assessment in which an incidence difference of 10/100,000 is considered acceptable. Further, assume that in this assessment the exposure intensity is high, the dose is medium, and the dose-response parameter beta is medium. Inspection of Figure 6-3 indicates that the difference in predicted incidence will be <10/100,000 (in the top 3x3 grid, the square in the top row, middle column is gray). On the other hand, if the exposure intensity is medium, the dose is high, and the dose-response parameter beta is low, Figure 6-3 indicates that the difference in predicted incidence will not be <10/100,000 (in the top 3x3 grid, the circle in the middle row, right column is white).

For the first example provided above it would be reasonable to employ a static risk model, since the difference in predicted incidence is less than the threshold level (<10/100,000). In the second example, a dynamic model would be more appropriate because either person to person transmission or immunity impacts the assessment sufficiently such that the predicted

---

[14]The threshold levels of tolerable incidence difference between models were selected based on the available scientific literature and best professional judgment. The WERF Project Subcommittee and Risk Advisory Committee reviewed these values and agreed that they are a reasonable, representative range for decision-making purposes.

levels of incidence in the static and dynamic models differ by more than the predetermined threshold level (10/100,000).

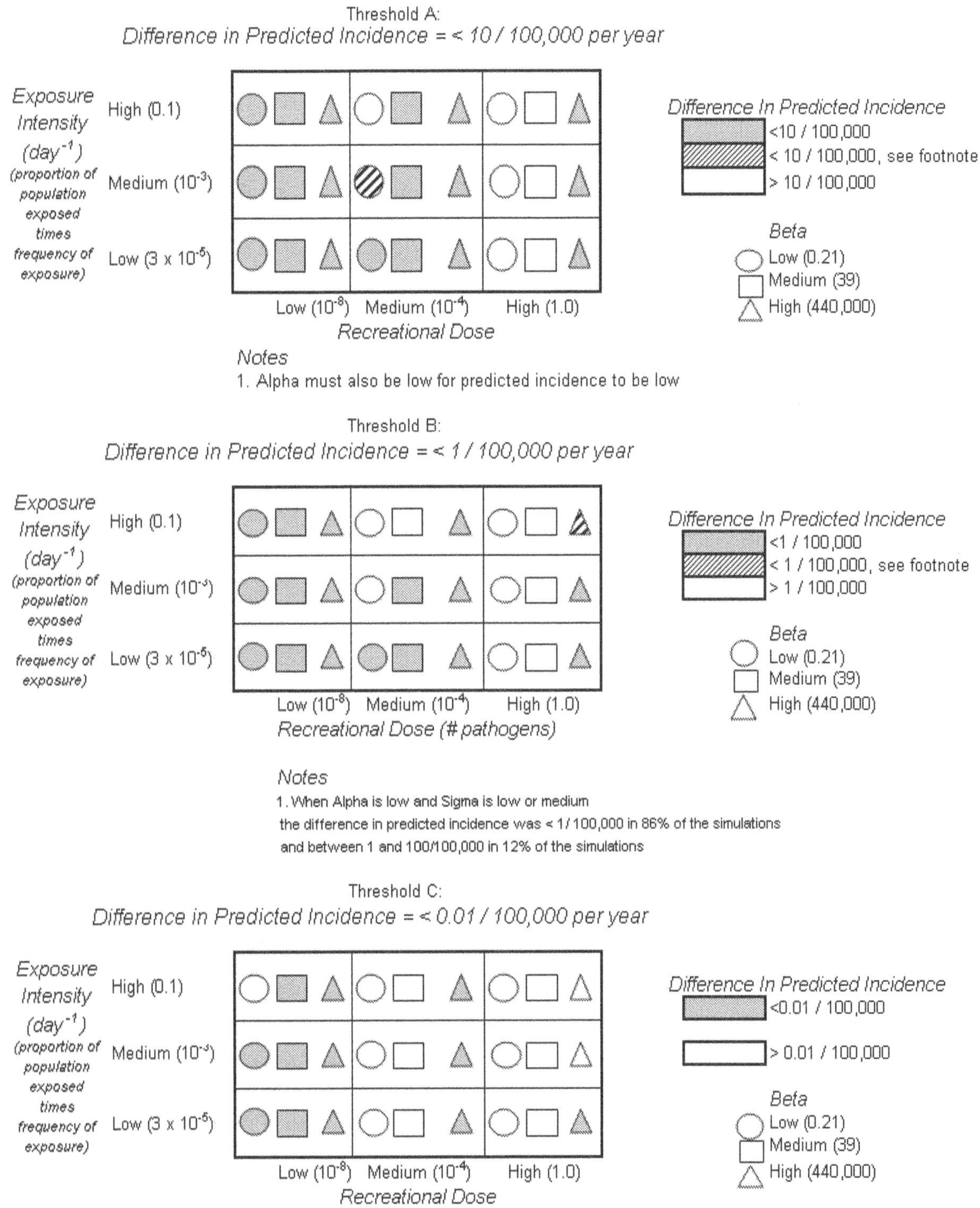

Figure 6-3. Graphical Summary of Microbial Risk Assessment Model Evaluation.

If it is assumed that typical human exposures to stormwater would not be greater than exposures to recreational waters, the results of the previous WERF research can be used to help identify an appropriate risk assessment methodology for exposure to microorganisms in stormwater. The components that need to be considered are the exposure intensity, the dose of pathogens to which individuals would be exposed and the acceptance threshold for difference between the different model results.

The exposure intensity is defined as the product of the proportion of the population exposed and the frequency of exposure (for example a once per month exposure would have a 1/30 frequency of exposure). For exposures to stormwater it is reasonable to expect that in most cases throughout the U.S., the proportion of the population exposed and the frequency of exposure would both be quite low (for example less than 1/1000th of the population exposed less than once per month), yielding a low exposure intensity (bottom row of 3x3 grids in Figure 8-3).

As discussed in Chapter 2.0, a limited amount of data is available to characterize pathogen concentrations in stormwater. Using assumed volumes of ingestion similar to those assumed for recreational waters and the median concentration values of pathogens in stormwater that are available (Table 2-4), it is possible to estimate on a screening level, exposure doses for stormwater exposures (following the approach from recreational water investigations, that ingestion is the most important route of exposure). Based on these assumptions, stormwater exposures could result in virus and parasite exposures that fall in the “medium” or “high” exposure categories, and bacteria exposures that fall into the “high” exposure category. (Note that “beta” for viruses of concern is typically low, “medium” for parasites, and “high” for bacteria except *Shigella* whose “beta” is “medium”).

Selecting an acceptance threshold for difference between the different model results is the final requisite step in the model selection process. Typically a low acceptance threshold would be selected in conditions in which there is little room for error. A low acceptance threshold between model results would also indicate a desire for high precision along in the resulting analysis. On the contrary a higher acceptance threshold for difference between the different model results would be selected for a screening level analysis or another type of analysis in which great precision is not necessary to make appropriate inferences. Given the lack of data that are currently available, it is assumed that for the purposes of this investigation, that the lowest threshold for difference between the different model results is not necessary and that appropriate inferences can be made based on other acceptance thresholds.

Based the inferences and assumptions described above and the model evaluation guide presented in Figure 6-3 it appears that a static MRA methodology may be most appropriate for stormwater risk assessments at this point in time. (Refer to the top and/or middle grids, bottom row, and middle and/or right column). In most cases, for the specified combinations, a static model would result in differences that are lower than the acceptance criteria. In cases where the average dose of viruses or parasites fall into the “high” category (average of ~1 pathogen ingested) the acceptance criteria of a difference of less than 1 or 10 illnesses per 100,000 people per year may not be met.

Given the limited amount of data available to characterize concentrations of pathogens in stormwater, it is recommended that a static model be used initially for this investigation. Once additional data are available, it is recommended that the issue of model selection be revisited.

CHAPTER 7.0

SCIENCE-DIRECTED FIELD PROGRAM FOR DETERMINING RISK ASSOCIATED WITH EXPOSURE TO PATHOGENS IN STORMWATER

1. Which waterborne pathogens pose the greatest risk to human health and what are the concentrations of these pathogens and indicator organisms observed in stormwater?
2. Are there defensible relations between indicator organisms and pathogens of public health concern in stormwater?
3. What is the fate of stormwater microorganisms (i.e., pathogens and indicator organisms) in the environment?
4. What microbial source tracking (MST) techniques are used to identify sources and assess potential public health risk?
5. What are the most effective stormwater microorganism control technologies and what are the estimated costs?
6. What methods have been historically used to quantitatively characterize potential human health risk associated with exposure to waterborne pathogens and what is the most appropriate method for stormwater?
7. What critical data are needed to enable the assessment of risk associated with exposure to pathogens in stormwater and what is the recommended approach?

7.1 Introduction

The principal goal of this report is to provide insight relative to the potential human health risk associated with exposure to pathogens in stormwater. Risk associated with these organisms can be quantitatively characterized through epidemiology study results or if data are available to model risk. The literature review conducted in this report revealed that an insufficient number of epidemiology studies relate human illness to stormwater exposure, and that the data available to parameterize human health risk models are not robust.

This chapter develops a prioritized, phased approach for acquiring the data necessary to characterize the range of potential human health risks associated with exposure to microorganisms in stormwater (as defined in Chapter 1.0). The proposed plan of action emphasizes development of methods and sampling schemes to overcome key inconsistencies in collection of microorganisms in stormwater and analysis that made the few available studies of concentrations of microorganisms in stormwater insufficient for performing a defensible risk assessment

7.2 Critical Data Gaps for Risk Assessment

Based on the information presented in Chapter 6.0, a static risk model would require the following inputs: pathogen concentration, volume of water ingested, dose-response relationship for the host-pathogen interaction, and a description of the population exposed. Review of the available literature reveals a striking lack of information regarding the two most important inputs to the risk model: pathogen concentrations and exposed populations. The few available studies of concentrations of microorganisms in stormwater were insufficient for performing a defensible risk assessment because:

1. many of the most important waterborne pathogens have not been quantified in stormwater;
2. enumeration methods used by researchers were disparate and did not account for particle-bound pathogens; and,
3. the sampling time during the storm was not standardized, nor known to have occurred during a period when there was human exposure.

Each of these issues is described in more detail below.

7.2.1 Pathogens Enumerated

The concentrations of some of the most important waterborne pathogens are missing from Table 2-4, which summarizes the available data on pathogens in stormwater. According to Mead et al. (1999), analysis of foodborne and non-foodborne illness in the U.S., enteric viruses (primarily transmitted through non-food routes) caused almost 80% of these illnesses. The most important of these enteric viruses are noroviruses. Based on available data, there have been no studies of noroviruses in stormwater. Rotavirus and enterovirus are also important etiological agents of waterborne disease, yet only one and five studies, respectively, quantified these organisms. The protozoans *Cryptosporidium* and *Giardia* are also important agents of waterborne disease, yet only 11 studies have enumerated them. The bacterial pathogens *Salmonella*, *Shigella*, and *Campylobacter* together are the third most important etiologies of waterborne illness, yet only four studies worldwide have enumerated *Salmonella* in stormwater, and no studies have quantified *Shigella* and *Campylobacter* in stormwater. Other epidemiologically important waterborne pathogens include toxigenic *E. coli*, astrovirus, and hepatitis A. The project team is not aware of any studies that have been conducted on these organisms, except for one on hepatitis A by Jiang and Chu (2004).

7.2.2 Enumeration Methods

Methods used to quantify pathogens in stormwater have differed concerning actual enumeration method and accounting for particle-associated organisms. There are U.S. EPA or standard methods for enumerating fecal indicator organisms and some pathogens, but there are no standard methods for enumerating many bacterial and viral pathogens. Although nucleic-acid based viral assays have become popular over the last decade because of advances made with quantitative-PCR (qPCR), researchers have not reached a consensus as to how concentrations determined by qPCR relate to concentrations of infectious pathogens. Choi and Jiang (2005) report that their measurements of density of genome copies of adenovirus measured using qPCR correlate poorly to densities of infectious adenovirus obtained using two continuous cell lines. The peer-reviewed literature has not extensively addressed the presence of particles in stormwater and their effects on pathogen enumeration. In fact, few studies mention particles and their influence on pathogen enumeration in their samples. A study aimed at accurately enumerating pathogen concentrations in stormwater must consistently account for suspended particle-associated pathogens in the analyses.

7.2.3 Sampling Time During a Storm

One of the major problems in interpreting the data presented in Table 2-3 is the lack of standardization in the timing of sample collection during storm events. The concentration of pollutants in storm runoff may change drastically over the duration of a storm (Bertrand-Krajewski, 1998). In addition, depending on the antecedent dry weather period, and the rainfall intensity and duration, the concentration of pollutant may vary from storm to storm. In general, the variation of pollutant levels within and between storms is not predictable given the current state of knowledge. As shown in Table 2-3, few studies describe the timing of samples relative to the onset of a storm or to antecedent rainfall. The studies also did not indicate whether humans were exposed to stormwater during the storm that they sampled. Hence, exposure frequency during sampling was not quantified.

7.3 Overview of Recommended Plan of Action

The recommended plan of action is a prioritized, phased approach to acquire the data necessary to characterize the range of risks associated with exposure to microorganisms in stormwater. This plan of action emphasizes development of methods and sampling schemes to overcome the key inconsistencies in the available studies. This section provides an overview of the recommended plan of action. More details follow in subsequent sections.

7.3.1 Phase 1a: Interim Guidance and Management Prioritization

It is recognized that microorganisms in stormwater currently represent a concern to many municipalities in the U.S. Further, if one applies existing recreational water criteria to waterbodies dominated non-point source (NPS) discharges, the implicit assumption would be that (roughly) the same indicator-to-pathogen relation is present in NPS-impacted water as was present during the epidemiological investigations for recreational waters. Based on the review of the literature, it appears that the available data are insufficient to support that assumption. Thus, a prioritized interim guidance for management prioritization is recommended.

7.3.2 Phase 1b: Methods Development

The concentration of pathogens in stormwater is a key input to a risk-based model. However, few of the pathogens of public health concern likely to be most important in stormwater are measurable at levels that can provide insights necessary to make critical management decisions. The first step in the assessment of risk associated with exposure to storm runoff microorganisms is investing in research to improve techniques for enumerating pathogens in stormwater.

7.3.3 Phase 2: Pilot Monitoring Program and Risk Assessment in a Single Watershed

Once methods are available to characterize concentrations of pathogens of public health concern at critical levels, a screening-level risk assessment could be conducted. Therefore, it is recommended that a pilot monitoring program be conducted in a single watershed, followed by a risk assessment investigation. This pilot-scale investigation will encompass many of the aspects of a national monitoring program (land use, sampling location identification, inter- and intra-storm variability, human exposure, etc.), but will represent a modest investment; the lessons learned will increase the likelihood of gathering useful data in a subsequent full-scale investigation.

7.3.4 Phase 3: National Stormwater Monitoring Program and Associated Risk Assessment

The overall goals of a national stormwater monitoring program will be to quantify 1) the concentration of target infectious/viable pathogens in stormwater at the "end of the pipe" and 2) the number of individuals exposed to stormwater in receiving waters for various types or classes of stormwater. Table 7-1 shows the key objectives of such a national field program.

Table 7-1. Key Objectives of a National Stormwater Monitoring Program Aimed at Risk Assessment.

1. Determine the range in concentrations of key pathogens at the "end of the pipe" of a stormwater outfall, watershed outlet, or conveyance adjacent to receiving waters.
2. Relate the presence and concentrations of pathogens and fecal indicators in stormwater to stormwater types so that priority may be given to specific types of stormwater for further study.
3. Quantify the types of exposures and number of individuals exposed to stormwater in the adjacent receiving waterbody.
4. Relate exposure types and numbers of individuals exposed to stormwater in receiving waters to stormwater types to give priority to specific types of stormwater for further study.

Once the national stormwater monitoring program is completed, the next goal of this component of the recommended plan of action will be to develop a generalizable risk model. The model should relate risk to easily quantifiable watershed characteristics so that specific watershed types may receive high priority for future study and monitoring. Use of data collected during previous phases in this manner could help direct management actions, inform the potential utility of future epidemiology studies, and inform regulatory limits for microorganisms in stormwater.

7.4 Phase 1a: Interim Guidance and Management Prioritization

Microorganisms in stormwater currently represent a perceived or actual concern to many municipalities in the U.S. Based on the information presented in this report, it should be clear that quantitatively characterizing the human health risk associated with exposure to microorganisms in stormwater is currently impractical. Subsequent sections of this report present a recommended approach to acquiring the data necessary to characterize the range of risks associated with exposure to microorganisms in stormwater.

Although quantitative characterization of the human health risk associated with exposure to microorganisms in stormwater is not currently practical, it would be possible to develop an interim guidance plan that could inform management prioritization considerations. Such interim guidance could be based on qualitative information and expected levels of exposure. For example, Soller et al. (2003) suggested that a qualitative approach may facilitate risk management related to recreational water beneficial use impairment. In that case, data from a sanitary survey, relevant water quality data, and specific knowledge about microbiological sources are used along with water quality monitoring and exposure data to derive a relative level of public health concern associated with recreational activities. Soller et al. (2003) suggested that management could act based on the prioritized level of public health concern.

Similarly, Soller et al. (2003) suggested a qualitative or semi-quantitative approach to prioritize and evaluate the potential human health impact of combined sewer overflows (CSOs) and sanitary sewer overflows (SSOs), and to assist in determining where public resources could be most effectively used to implement control or response strategies aimed at minimizing those potential impacts. The researchers derived a series of lines that qualitatively separate CSO and SSO scenarios above the acceptable level of risk from those near or below the acceptable level of risk, for decreasing levels of exposure (Figure 7-1). The concepts set forth in Figure 7-1 are intuitively consistent with public health protection, conceptually straightforward, and potentially applicable to stormwater exposures. For example, consider a sewer overflow event that results in little dilution and is located near an area that supports a light level of recreational activity (Figure 7-1, Example 1). Inspection of Figure 7-1 indicates that such a scenario would likely be cause for public health concern should exposure occur, and thus may require some action (e.g., beach closing or implementation of another control measure to reduce exposure). Similarly, if a overflow event with a substantial amount of dilution (e.g., >100:1) occurred at the same beach, the level of public health concern may be somewhat diminished (Figure 7-1, Example 2). As a final example, consider an SSO (little to no dilution) located in an area where there likely to be only incidental contact by very few people (Figure 7-1, Example 3). Inspection of Figure 7-1 indicates that such an event would likely not be cause for public health concern. If, on the other hand, the SSO occurred in an area where an activity resulted in a higher level of exposure, it may warrant more public health concern.

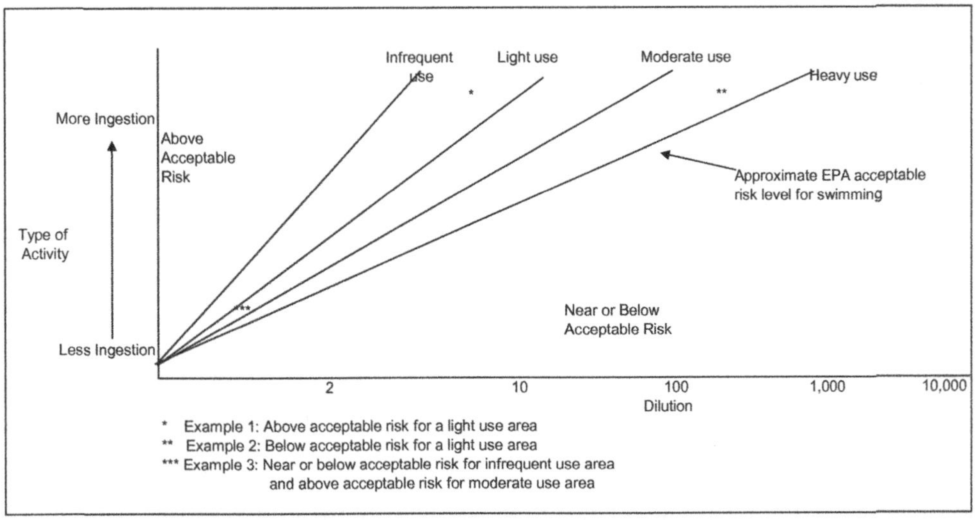

Figure 7-1. Conceptual Relation between Risk Associated with Sewer Overflows and Relative Number of Exposed Individuals. Adapted with permission from Soller, J., Turbow, D., Olivieri, A. Characterization of the Potential Adverse Human Health Effects Associated with Combined Sewer Overflows and Sanitary Sewer Overflows. Report for the Association of Metropolitan Sewerage Agencies, EOA Inc: Oakland, CA, 2003.

The project team recommends that a reasonable and appropriate first step in the assessment of risk associated with exposure to storm runoff microorganisms is the development of an interim guidance plan that could provide management prioritization considerations. The emphasis of this recommended interim guidance plan should be the use of available information to facilitate risk management and resource allocation prioritization decisions.

7.5 Phase 1b: Methods Development

7.5.1 Background

7.5.1.1 Available Pathogen Enumeration Methods

There are a handful of methods for determining concentrations of pathogens in stormwater. These include nucleic-acid based, culture-based, microscopy-based, and hybrid detection techniques. An overview of these methods follows.

Schroeder et al. (2002) and Rajal et al. (2005) at the University of California, Davis (U.C. Davis) have undertaken a large-scale effort to develop methods for the enumeration of pathogens in stormwater in cooperation with the California Department of Transportation (Caltrans). Their tool of choice is real-time, quantitative polymerase chain reaction (qPCR) with fluorescently labeled internal probes. qPCR is a fast assay for measuring concentrations of pathogens in environmental media. Assays can be run in three hours and multiple pathogens can be assayed at the same time using multiple primers and probes. Despite these benefits, there are a number of drawbacks to using qPCR to enumerate pathogens in environmental waters generally, and stormwater in particular.

Quantitative PCR (qPCR) biochemistry is very sensitive to impurities, such as heavy metals and organic matter (including humic acids and fats). These impurities are likely to be present in high concentrations in stormwater. Schroeder et al. (2002) found that the concentration of suspended solids in sieved raw stormwater samples played a statistically significant role in production of inhibitory compounds. Even though sample concentration can help improve the detection limit of pathogens in environmental media, the presence of inhibitors can force scientists to dilute the nucleic acid extract. However, dilution may bring the extract below method detection limits, rendering the concentration useless.

The detection limit of qPCR is high relative to epidemiological significant concentrations, due to the presence of impurities in the media. According to Schroeder et al. (2002), "…inhibition and interferences associated with environmental samples can result in unacceptably high [PCR] detection limits". Loge et al. (2002) pointed out that PCR detection limits are currently not low enough to provide information relative to risk assessment, and the study of Schroeder et al. shows that this very well may be the case, particularly for viruses. The performance of qPCR relative to other detection techniques in detecting epidemiological significant concentrations of pathogens will be discussed below.

The final drawback of qPCR is that it measures only the concentration of genomic material, or gene copies. The presence and concentration of gDNA or gRNA do not necessarily correlate to the presence of infectious or viable organisms. The work of Choi and Jiang (2005), who measured high concentrations of adenoviruses in urban rivers using qPCR, but found very low infectivity when the same water samples were assayed using tissue culture with two different cell lines, illustrates this well. Choi and Jiang's results suggest that adenovirus gDNA may persist outside the virion or in damaged virions that cannot infect host cells. Because gDNA is also the qPCR target for bacteria and protozoa, the lack of correlation between qPCR- and culture-derived concentrations for adenovirus might extend to other DNA viruses, bacteria, and protozoa. Choi and Jiang's results are not necessarily extendable to RNA viruses. RNA is more labile in the environment than DNA, so presumably dissolved RNA would not persist for an extended time outside the virion. It is not known whether qPCR results from RNA viruses correlate to concentrations of infectious virions. One possibility for reducing the detection limits of non-infectious, non-viable bacteria or protozoa in stormwater is to focus qPCR assays on messenger RNA (mRNA) of bacteria and protozoa rather than gDNA, thus taking advantage of the labile nature of RNA.

Schroeder et al. (2002) explored the use of PCR for detecting pathogens in stormwater in California. Given all the impurities in stormwater, the lowest detection limits in stormwater described by Schroeder et al. were approximately 170 viruses per L, and 150 bacteria or per L. No detection limit for protozoa was reported by Schroeder et al. (2002), however, we have applied the detection limit they reported for bacteria to protozoa. In reality, protozoa have more gDNA than bacteria, and are larger, thus their detection limit could be lower. The worst detection limit achieved by Schroeder et al. for viruses was 30,500 viruses per L.

Culture-based assays are the only pathogen assays that allow enumeration of viable organisms that are likely capable of infecting a human. For pathogenic bacteria, culture-based assays require selective enrichment and growth on specific agars, followed by biochemical and/or molecular confirmation assays. Culture-based assays for detection of bacteria are open to criticism because they cannot enumerate viable but unculturable pathogens. Pommepuy et al. (1996) showed that enteropathogenic *E. coli* enter a viable but non-culturable (VBNC) state but remain pathogenic. Detection limits for bacteria in water are typically approximately 0.1 to 1 per 100 mL when membrane filtration is used. The detection limit could theoretically be lowered by using a membrane filtration-most probable number (MPN) method, like that described for *Salmonella* in U.S. EPA method 1682.

For infectious viruses, culture-based assays require membrane filtration, tangential or vortex flow filtration to concentrate viruses from water samples. These techniques rely on electrostatic interactions between virus and filter for collection or size-selective concentration. Mammalian cell lines are required to perform plaque-forming or presence/absence assays. Table 9-1 shows the mammalian cell lines used for the enumeration of viruses likely to be important in stormwater. Once a water sample is concentrated, host cell cultures are inoculated with sample. After two to four weeks, if the cell culture shows signs of cytopathic effect (CPE) the sample is determined to be positive for human virus. Norovirus, potentially the most important virus to assay for because it is responsible for the most gastrointestinal illnesses in the U.S. (Mead et al. 1999), cannot be propagated in a cell line. CaCO2 can be used to culture numerous viruses (except for norovirus), and it has been used by some researchers as a "catch-all" cell line for detecting viruses in the environment (e.g., Chapron et al., 2000). The lowest concentrations for enteroviruses using cell culture reported in the literature described in Chapter 3.0 were 0.00175 MPN per L (Rose et al., 1987), 0.2 MPN per L (Reynolds et al., 1998), and 0.01 MPN per L (Lipp et al., 2001). The drawbacks to using cell culture assays to detect human viruses are summarized by Griffin et al. (2003), and include 1) long duration, 2) low sensitivity, 3) high cost, and 4) the process is labor intensive. Table 7-2 shows potential virus targets for the stormwater field program and cell lines (if applicable) that can be used to assay for infectivity. Note that there is no known cell line for the propagation of norovirus.

Table 7-2. Potential Virus Targets and Cell Lines.

Virus	Continuous Cell Line
Enterovirus	BGM, HeLa, $CaCO_2$
Hepatitis A	$CaCO_2$, FRhK-4, BGM, MRC-5
Rotavirus	$CaCO_2$, MA104
Norovirus	**NO KNOWN CELL LINE**

Concentration is also a critical step for enumeration of protozoa by cell culture. *Cryptosporidium* can be propagated and enumerated in HCT-8 and CaCO2 cell lines (Rochelle et al., 1997; Schets et al., 2005). Schets et al. (2005) reported detection limits on the order of 0.01 oocysts per L for *Cryptosporidium* in surface waters. *Giardia* can also be propagated in cell culture (Bowie et al., 1988), but few studies have applied these to environmental waters.

For pathogenic protozoa, the U.S. EPA methods for enumeration require the use of fluorescently labeled antibody probes for microscopic enumeration (U.S. EPA methods 1622 and 1623). After concentration, antigen-specific antibodies that are fluorescently labeled are applied to the water sample, which is subsequently screened using a microscope. For human viruses, a similar assay can be used where antigen-specific, labeled antibodies are used to detect viruses in a water sample. Unfortunately, extremely high concentrations of virus (approximately 10 million viruses per mL) must be present for the method to work. Therefore, it is not useful for water samples, unless used in conjunction with a cell culture assay (Griffin et al., 2003).

Recent work has investigated an integrated cell-culture PCR (ICC-PCR) assay for human viruses and protozoa (Reynolds et al., 1996, Rochelle et al., 1997, Griffin et al., 2003, Lee and Jeong, 2004). ICC-PCR has the advantages of both cell culture (detects infectious pathogens) and PCR (fast). Briefly, a host cell line is seeded with an environmental sample. After 24-48 h, before CPE can be observed, an aliquot of cell culture is extracted and analyzed with PCR for the presence of virus or protozoa. Such assays are relatively new and are in development with the goal of providing rapid detection of infectious human viruses in aqueous matrices. Reynolds et al. (1997) have reported detection limits for ICC-PCR for enterovirus at 0.001 MPN/liter.

7.5.1.2 Detection Limits Considerations

From the list of pathogens identified by U.S. EPA during their research related to combined and sanitary sewer overflows (U.S. EPA, 2002a), CDC's estimated disease burden in the U.S., and the relatively large base of knowledge from epidemiological studies conducted in recreational waters[14], the waterborne pathogens of public health concern likely to be most

[14] The most frequent and consistent adverse health outcome associated with exposure to fecally contaminated recreational water is gastroenteritis (or GI) illness (WHO, 2003; Wade et al., 2003). In addition, a cause-effect

important relative to applying recreational criteria to stormwater are: noroviruses, enteroviruses, hepatitis A, rotavirus (human viruses), *Cryptosporidium parvum, Giardia lamblia* (protozoa), *Salmonella* spp., *Shigella* spp., *Campylobacter* spp. and toxigenic *E. coli* (bacteria). These are referred to as the target pathogens herein.

As described in Chapter 6.0, the concentration of pathogens in stormwater is a central input to a risk-based model. Using the concentration, in conjunction with the volume of ingested water and the dose-response relationship, one can calculate the risk of illness or disease for a susceptible individual. Using a static risk model, it is possible to "back calculate" the concentrations of pathogens we need to be able to measure. Such calculations can help dictate which detection technique is acceptable for enumerating pathogens, or the need for further method refinement.

Table 7-3 shows the concentrations of rotavirus that would need to be measured in stormwater to determine specific risks (columns) based on an assumed volume of ingested water (rows). In Table 7-3, the dose-response function for rotavirus is shown at the top of the table along with the dose-response variables. It is typically assumed that during full-contact recreation, about 100 mL of water is ingested. If we wish to detect tolerable risks greater than or equal to 0.008 (8/1000 become diseased), the tolerable risk basis for REC-1 freshwater microbial standards15, then we would need to be able to measure rotavirus at levels greater than or equal to 0.14 rotavirus per L of water. **Bolded** numbers highlight the concentrations that would be detectable with the best-case scenario detection limit of PCR determined by Schroeder et al. (2002).

Table 7-3. Concentrations of Rotavirus Resulting in Designated Tolerable Risks.

Pathogen	rotavirus				
Dose Response function	$PI(d) = 1-[1+(conc*vol)/\beta]^{-\alpha}$				
Dose Response parameter(s)	0.2531	0.4265			
	Tolerable Level of Risk of Infection				
Volume Ingested (L)	0.0001	0.001	0.008	0.01	0.1
0.5	0.0003	0.003	0.028	0.03	0.4
0.1	0.002	0.02	0.14	0.17	2.2
0.05	0.003	0.03	0.3	0.35	4.4
0.01	0.02	0.2	1.4	1.7	22
0.001	0.17	1.7	13.8	17.3	**220**
0.0001	1.7	17	138	**173**	**2202**

Table 7-4 shows a similar exercise, performed for non-typhi *Salmonella*. Under the same criteria of tolerable risk and volume ingested, we would need to be able to measure non-typhi

---

relationship between fecal or bather-derived pollution and acute febrile respiratory illness (AFRI) has also been reported (Fleisher et al., 1996). More specifically, a significant concentration-response relationship between AFRI and enterococci was established. However, when compared with GI illness, the probabilities of contracting AFRI are generally lower and the threshold at which illness is observed is higher.

[15] U.S. EPA (1986). *Ambient water quality criteria for bacteria*, EPA440/5-84-002.

Salmonella in concentrations equal to or greater than 751 *Salmonella* per L.

Finally, Table 7-5 shows similar results for the TAMU isolate of *Cryptosporidium*. Inspection of Table 7-5 indicates that we would need to be able to measure *Cryptosporidium* concentrations greater than or equal to 1.0 oocyst per L of stormwater to measure risks at the REC-1 level. As a reference level, the detection technique for pathogens should at least be able to detect pathogens at these levels. In both tables, **bolded** numbers highlight the concentrations that would be detectable with the best case scenario detection limit of PCR determined by Schroeder et al. (2002).

Table 7-4. Concentrations of *Salmonella* Resulting in Designated Tolerable Risks.

Pathogen	salmonella nontyphi				
Dose Response function	$PI(d) = 1-[1+(conc*vol)/\beta]^{-\alpha}$				
Dose Response parameter(s)	0.3126	2884			
	Tolerable Level of Risk of Infection				
Volume Ingested (L)	0.0001	0.001	0.008	0.01	0.1
0.5	1.8	18	**150**	**188**	**2312**
0.1	9.2	92	**751**	**942**	**11559**
0.05	18	**185**	**1501**	**1885**	**23118**
0.01	92	**925**	**7506**	**9423**	**115592**
0.001	**923**	**9245**	**75064**	**94230**	**1155916**
0.0001	**9228**	**92453**	**750638**	**942295**	**11559165**

Table 7-5. Concentrations of *Cryptosporidium* Resulting in Designated Tolerable Risks.

Pathogen	Cryptosporidium (TAMU isolate)				
Dose Response function	PI(d) = 1-exp(-dose/k)				
Dose Response parameter(s)	12.99				
	Tolerable Level of Risk of Infection				
Volume Ingested (L)	0.0001	0.001	0.008	0.01	0.1
0.5	0.0026	0.03	0.2	0.3	2.7
0.1	0.013	0.13	1.0	1.3	14
0.05	0.026	0.3	2.1	2.6	27
0.01	0.13	1.3	10.4	13.1	137
0.001	1.3	13.0	104	131	**1369**
0.0001	13.0	130	**1043**	**1306**	**13686**

The only studies targeted to specifically measure concentrations of pathogens in stormwater are those that use qPCR and PCR. The studies described by Schroeder et al. (2002) and Rajal et al. (2005) are currently on-going at U.C. Davis with funding and guidance from CalTrans. (See 7.5.1.1 above; see also Appendix D for a brief description of their methods.) Clean-up procedures such as magnetic capture hybidization (MCH, magnetic beads coated with streptavidin with biotin labeled oligonucleotide probes) are currently being developed with the goal of achieving lower detection limits. Improvements in detection limits between 8 and 2000 fold have been achieved using MCH with *Salmonella* (Thompson et al., 2006).

Despite potential for improvement, qPCR detection limits are presently significantly higher for virus and protozoa than the concentrations required to assess risk at levels comparable to risks targeted in the REC-1 standard (Loge et al., 2002) (Tables 7-2, 7-3, and 7-4). An additional concern regarding the use of nucleic-acid based detection techniques for measuring pathogen concentrations in stormwater is that the detection of genetic materials is not equivalent to the detection of infectious organisms. The dose-response relationships discussed in Chapter 8 and used to generate Tables 7-2 through 7-4 were developed by measuring the dose of infectious organisms and comparing it to specific health outcomes. Because qPCR does not necessarily enumerate infectious organisms, concentrations derived using this technique that are used as inputs to a risk assessment may overestimate risk.

7.5.1.3 Considering Particle-Associated Pathogens in Stormwater

Particle-associated pathogens pose a challenge to measuring pathogen concentrations and thus assessing risk. Pathogens, and especially viruses, can be particle reactive, as described in Chapter 3.0 (e.g. Gantzer et al., 1994; Searcy et al., 2005; Characklis et al., 2005). Numerous researchers have documented the result of particle-pathogen interactions: particle-associated pathogens (see Chapter 3.0). Thus, it is imperative to desorb pathogens from particles prior to enumeration in order to get an accurate representation of pathogen concentration. Figure 7-2 illustrates a potential problem that could arise when enumerating microorganisms adsorbed to particles. When the particle on the left (with four microbes sorbed to it) is assayed using an colony forming unit (CFU) or plaque forming unit (PFU) assay, the same number of colonies or plaques will arise as the particle on the right (with one bacteria on it); both will give 1 CFU or PFU. This result would lead to an underestimate of the risk associated with exposure to water containing these two types of particles. Thus, efforts to accurately enumerate the number of microbes sorbed to sediment is necessary; this requires a desorption step.

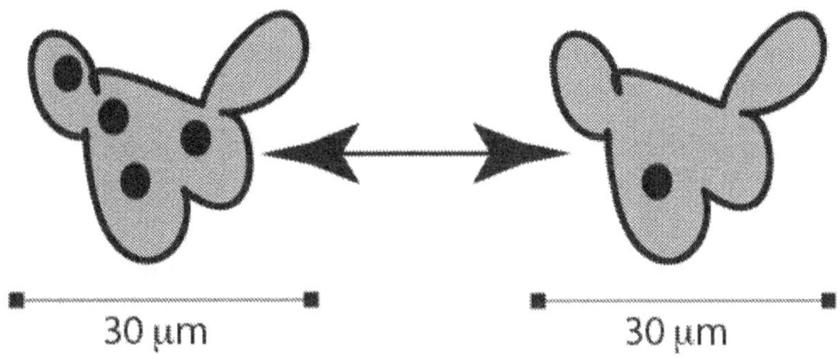

Figure 7-2. A 30 μm Particle that has 4 Microbes (black circles) (on right) or One Microbe (left) Sorbed to its Surface May Give the Same Result in a Culture-based Colony Forming Unit, or Plaque Forming Unit Assay.

Desorption of pathogens from particles is particularly challenging because pathogen attachment to particles may be irreversible. If desorption from suspended particles in a water sample is desired, then physical agitation (Craig et al., 2002; Borst and Selvakumar, 2003) is a possible method. Additional methods for desorbing pathogens from particles in sediment include: elution with high ionic strength buffers or high protein solutions (e.g., beef broth). Many researchers (Oshiro and Fujioka, 1995; Crabill et al., 1999; Davies and Bavor, 2000; Whitman and Nevers, 2004) eluted bacteria from sediments by shaking in phosphate-buffered saline. Ferguson et al. (2005) desorbed bacteria from sands by suspending sands in sodium metaphosphate and sonicating for 30 seconds. Roper and Marshall (1979), Buckley et al. (1998),

Desmarais et al. (2002) and Alm et al. (2003) used shaking in distilled water to desorb organisms from sediment. The most effective method for dislodging microbes from particles may depend on particle characteristics. For example, in a method comparison, Craig et al. (2002) compared manual shaking, a sonication bath, and a sonication probe in a medium of 0.1% peptone water to determine the best manner of desorbing bacteria from sediments. They found that two rounds of sonication were most effective at removing particles from sediments that had low clay content, while physically shaking sediment was most effective at desorbing bacteria from sediments with high clay content.

As outlined in Chapter 3.0, the few studies that have examined the proportion of microorganisms in stormwater associated with suspended particles have documented that up to 100% of them are particle-associated. These studies were limited in that they did not attempt to desorb pathogens from particles during enumeration. Because pathogens sorbed to particles could result in an underestimate of the true number of pathogens in a water sample (Figure 7-2) when assays are culture-based, it is imperative to develop methods for desorbing pathogens from suspended particles prior to enumeration.

7.5.1.4 Concentrating Water Samples for Viral and Protozoan Detection

All pathogen assays require small volumes of water. In the case of PCR detection assays, extremely small purified nucleic-acid samples are processed in the reaction tube (usually 10 µL). For tissue culture, volumes on the order of 1 mL are used. Because these organisms are typically present in low concentrations in environmental samples and are epidemiologically significant at low concentrations, concentrating water samples is a necessary step. Membrane filtration, filtration by large woven yarn filters, tangential filtration, vortex filtration, and chemical precipitation have all been used (Griffin et al. 2003; U.S. EPA method 1622 and 1623). While culture-based bacterial assays also require concentration, it is possible to work with fairly large volumes of water using membrane filtration or membrane filtration- MPN methods, so concentration is not as much as a concern.

Stormwater represents an especially challenging matrix for detection of pathogens because it is extremely "dirty" with high concentrations of autochthonous microorganisms, particles, metals, and organics. Rajal et al. (2005) have determined an optimal method for concentrating pathogens in stormwater for use with qPCR. They achieved concentrations of up to 1000X using a two-step tangential membrane filtration system with 50,000 MW cutoff membranes. Their protocol also involved using sieves to remove particles in excess of 38 µm prior to filtration, and centrifugation to separate viruses from bacteria (and, potentially, protozoa) prior to the second filtration step. If concentration, recovery, and amplification are 100% efficient, this allows detection of pathogens at concentrations of ~1/100 L. During pathogen concentration, PCR inhibitors are also concentrated. Hence, although this step potentially increases sensitivity of the assay, it may reduce the potential detection limit by increasing inhibition.

Concentration of water samples is a key step for pathogen enumeration in stormwater. Efforts should continue to find the most efficient way to concentrate samples to reduce detection limit. New approaches are needed for removing inhibitors that are also concentrated and may interfere with the enumeration assay.

7.5.2 Methods Development Recommendations

Given the information available in the literature review on methods for determining concentrations of pathogens in water, it appears that ICC-PCR is able to detect viruses at the lowest concentrations, thus would be well suited for the required detection limits outlined above for protozoa and viruses. In addition, this technique is able to enumerate infectious viruses and protozoa. However, researchers have not applied ICC-PCR to stormwater; it is still in the developmental stage. Thus, it is not feasible to use this method currently in a stormwater health assessment study. It is also not realistic to extend the detection limits for ICC-PCR in drinking waters (Reynolds et al. 1997) to stormwater because stormwater contains so many more impurities.

It is recommended that a first step in the assessment of risk associated with exposure to storm runoff microorganisms should be the investment in research to improve techniques for the enumeration of pathogens in stormwater. Method development on pathogen detection in stormwater should precede the completion of a national monitoring program to assess risk associated with stormwater exposure. Method development should consider the following:

1. Concentration of microbes in stormwater is key for enumeration, so sample concentration should be an integrated step in the enumeration method.
2. There is presently no cell line for propagating one of the most important pathogens that needs to be measured (norovirus) in storm water; thus detection of infectious viruses may not be possible for all viruses.
3. The U.S. Congress has mandated that the U.S. EPA develop rapid methods for measuring microbial contaminants in the near future; thus assays for microorganisms in stormwater should be rapid (on the order of hours).
4. New methods for detecting microorganisms in stormwater should be extendable to published dose-response relationships between pathogens and health outcomes. This specifically implies that the concentrations of pathogens detected should be in terms of infectious units.
5. New method development should carefully consider the large amount of research that is currently being conducted
6. The potential for particles to have multiple pathogens sorbed to their surface must be considered during the development phase. The protocol should incorporate techniques for desorbing microbes from particle surfaces. These might include blending, sonication, or other methods (Craig et al., 2002).
7. The methods must incorporate the determination of detection limits, measured in terms of the number of infectious organisms per L of stormwater. This may require the use of internal standards.

7.6 Phase 2: Pilot Monitoring Program and Risk Assessment in a Single Watershed

7.6.1 Issues to Be Addressed in Development of Monitoring Plan

7.6.1.1 Type of Stormwater

Land use, including the amount of impervious surface, heavily influences the quantity of microorganisms in stormwater (Burton and Pitt, 2001). Based on the review of the peer-reviewed literature (Chapter 2.0), researchers have not elucidated the precise relationship between land use and microorganisms in stormwater. Selmakumar and Borst (2006) showed in their study of an urban watershed that high-density residential stormwater had higher concentrations of microorganisms than did low-density residential and landscaped commercial areas. To assess risk of human illness from exposure to stormwater, one should study stormwaters from a variety of land uses, including residential, commercial, industrial, agricultural and forest/open space lands. These land uses represent broad categories that include a number of constituent classes from the National Land Cover Dataset (NLCD) (USGS, 1992), as shown in Table 7-6. Although investigations should not be limited to urban or suburban regions because microorganisms in stormwater may emanate from wild or domesticated animals (Rajal et al., 2005), the pilot project will focus on urban land uses as more individuals are likely to be exposed to these stormwaters.

Table 7-6. Constituent NLCD 1992 Classes Included in Broad Category Classifications for Land Use.

Broad Category	Constituent Classes
Industrial	Commercial/Industrial/Transportation
Commercial	High-density residential
Residential	Low-density residential
Agriculture	Hay/pasture, row crops, grasslands/herbaceous, orchards/vineyards, small grains
Forest	Deciduous forest, coniferous forest, mixed forest, woody wetlands, shrubland, herbaceous wetlands, bare rock, transitional, mines

USGS (1992)

Climate plays an important role in controlling concentrations of microorganisms in stormwater by controlling soil characteristics and modulating rainfall. In subtropical and tropical soils that are rich in organics, fecal indicator organisms can persist for extended periods of time and even grow (discussed in Chapter 3.0); it is unknown whether this is an issue for pathogens, but the potential for environmental reservoirs should not be ruled out. The porosity of the soil also may affect concentrations of microorganisms in stormwater by allowing percolation of a portion of the storm flow into the subsurface (Izbicki et al., 2004). Climate is an important driver for concentrations of microorganisms in stormwater, primarily by controlling the length of antecedent dry weather and the intensity and duration of rainfall (Soller et al., 2004a; Burton and Pitt, 2005). In a Mediterranean/arid climate, like that experienced in the western U.S., winter is the rainy season and the antecedent dry weather periods can be as long as six months. Thus, the first storm of the season may have significantly different concentrations of microorganisms than the last storm of the season. In subtropical climates like those in Florida and the southeast U.S., rainfall can occur throughout the year. In the continental climate of the northeast and Midwest U.S., rainfall occurs primarily in the spring and summer. Climate can also impact the types of vegetation present in the watershed, which in turn may affect the manner in which storm runoff constituents interact with the watershed surface before entering receiving waters where exposure can occur. Table 7-7 provides example climates and corresponding U.S. EPA Rain Zones in the U.S. The pilot project should take place in a watershed with elevated exposure to stormwater discharges or high expected pathogen densities in stormwater, or both.

Table 7-7. Example Climates of the U.S. and Associated U.S. EPA Rain Zones that Should be Considered When Developing a National Field Program Studying Microorganisms in Stormwater.

Climate	U.S. EPA Rain Zone(s)	Rainfall Characteristics	Representative States
Arid Southwest/West	5,6	Rainfall in winter; long antecedent dry weather	California, Arizona, New Mexico, Nevada, Utah, Idaho, Montana, Colorado, Texas
Continental Climate/ Midwest & Northeast	1,2,4	Late spring and early summer; follows snowmelt	Illinois, Minnesota, Wisconsin, Michigan, Indiana, Missouri, Washington DC, Maryland, New Jersey, Ohio
Subtropical/Tropical	3	Year round rainfall	Hawaii, Florida

Previous research has shown that concentrations of microbial indicators (e.g., fecal coliforms) in stormwater are generally variable, but elevated throughout storm events (Maestre and Pitt, 2005). With regard to risk assessment, however, the resulting stormwater discharge for a particular storm may be divided into several types, including 1) initial discharge attributable to rainfall (beginning of storm), 2) highest concentration in stormwater during a rain event, and 3) discharge after a rain has ceased. Stormwater discharged during the beginning of the rain and after the rain event are important for a risk assessment because these represent waters to which exposure is likely to be high (individuals are still in the water when the rain starts or have returned to the water after the rain has ceased). It is also important to characterize the highest concentration of pathogens during a storm as this is when the potential dose for an exposure event is highest. Regardless, the goal of sampling throughout the storm event and during multiple storms is to characterize the concentrations of microorganisms in stormwater. Unfortunately, there is presently no way to predict, a priori, when the concentrations of pathogens will be the highest during a storm. Concentrations of contaminants in stormwater are non-steady, and any patterns observed are likely to be site and storm specific. In fact, it is estimated that in a particular watershed, storm patterns can only be elucidated after monitoring of approximately 30 storms (R. Pitt, personal communication, 2006). Consequently, sampling will need to continue throughout the storm at regular intervals, unless information is known a priori about intra- and inter-storm microorganism density variability.

7.6.1.2 Types of Exposures to Stormwater

Representative recreational exposures to stormwater include full-contact water sports (e.g. swimming, surfing) and limited- to non-contact water activities (kayaking, boating, fishing, wading). Participation in full-contact water sports will result in the ingestion of greater volumes of water than limited contact sports. Thus, full-contact might be viewed as riskier than limited-contact recreation for a given concentration of microorganisms in stormwater. However, full- and limited-contact exposures could lead to similar risks if they occur in stormwaters containing relatively low and high concentrations of pathogens, respectively.

Climate influences potential exposures to stormwater. In much of the arid southwest, Mediterranean California, and the maritime climate of the Pacific Northwest, water recreation at beaches or other water access points takes place primarily during the summer, which also coincides with the dry season. In the southeast subtropical U.S., rain and exposure occur year round, and is especially prevalent during the summer. In the continental climate of the U.S Midwest and northeast, rain occurs during seasons when the temperature is above freezing (spring through fall). Exposure to stormwater is likely to be greater in regions where rainfall is prevalent in the summer. This does not mean, however, that the risk will also be higher in places with year-round or primarily summer rainfall. Dose and exposure determine risk. For example, stormwater in the West may have greater concentrations of pathogens because of the long antecedent dry weather periods. Despite lower exposure frequencies during the winter rainy season, the population's risk may be equivalent to a location with high exposure frequency and a low concentration of pathogens.

7.6.1.3 Pathogens to Be Enumerated

As described previously, the target pathogens are: norovirus, enterovirus, hepatitis A, and rotavirus (human viruses); *Cryptosporidium parvum*, and *Giardia lamblia* (protozoa); and *Salmonella* spp., *Shigella* spp., *Campylobacter* spp., *Pseudomonas aeruginosa*, *Staphylococcus aureus*, and toxigenic *E. coli* (bacteria).

In addition to these pathogens, the following indicators are also important targets for enumeration during stormwater risk assessment studies: *Enterococcus*, *E. coli*, total coliform, coliphage (F+ and somatic), and Bacteroides (both total and human-specific). While few studies have actually found a relationship between indicators and pathogens in stormwater (Chapter 2.0), it is imperative to explore the relationship further because the potential benefits of identifying a easily measurable indicator are great. In addition, because the concentrations of indicators are likely to be much higher than pathogens, the presence of indicators in unusually high or low concentrations may suggest the need for further or reduced study, respectively, of a specific stormwater. Note that the sources of human pathogens in stormwater are likely to include domestic or wild animals (Rajal et al., 2005), so direct relationships between human indicators and pathogens may not be found.

7.6.2 Recommended Outline for Pilot Monitoring Program

7.6.2.1 Watersheds to Be Monitored

Based on our review of the literature on watershed factors that affect stormwater and our professional opinion regarding watershed factors that influence exposure intensity and type, we expect land use and climate to have the greatest influence on risk from exposure to microorganisms in stormwater. As stated earlier, a study objective of a national monitoring program will be to relate the range of risks from stormwater exposure to climate and land use (Table 7-1). This could potentially allow the development of a relation between risk and climate/land, to identify priority locations for further research. In addition, such a relationship can help develop qualitative information models about risk for particular watersheds (Soller et al., 2006).

A summary of the most important land use/climate combinations is shown in Table 7-7. The land uses shown (columns) are identified as the dominant land use within the watershed, although it is unusual to find a single land use watershed. Each row shows a climate with a representative state within the U.S. where a study could potentially take place. Testing of all combinations of land use and climate would require a suite of studies at 25 study sites. In this phased pilot study, the project team proposes to examine a single watershed prior to determining the total number of watersheds needed for a national risk assessment. Specifically, it is recommended that the phased pilot study occur in a watershed where pathogen concentrations and exposure are expected to be high. This could either be a location in Southern California where surfers frequently are exposed to stormwater during the winter rainy season, and the presence of pathogens in stormwater has been documented in the peer-reviewed literature. Alternatively, one could conduct the pilot study in a location in the subtropical Southeast where rainfall occurs during the summer, when recreation is likely to be high, and pathogens have been detected after or during rainy weather. In Table 7-8, Xs denote possible locations for the pilot study. The pilot study should initially be focused on urban land uses as it is more likely that recreation uses are higher in heavily populated urban watersheds. Table 7-8 assumes that land use and climate are the main drivers for concentrations of microorganisms in stormwater. Each column represents a dominant land use. Each row is a particular climate in the U.S.; in parentheses is a suggested state where the directed study might take place.

Table 7-8. Various Types of Land Use and Climate Combinations Present in the U.S.

Climate/Dominant Land Use (Location in the U.S.)	Industrial	Commercial	Residential	Agricultural	Forested
Arid Southwest (California)	X	X	X		
Continental Climate of Midwest (Great Lake States)					
Continental Climate of Northeast (New Jersey)					
Subtropical Southeast (Florida)	X	X	X		

Watersheds included in the field program should be selected so there is limited interference with separate sewer overflows and sanitary sewer cross connections. In addition, it is recommended that priority be given to watersheds where previous work has characterized stormwater pollutants.

7.6.2.2 Sampling Location

The precise sampling point (Figure 7-3) for the pilot study should be chosen to target the outfall of the storm sewer/ stormwater conveyance (X) that discharges into receiving waters.

While this sampling point does not necessarily represent a location where exposure actually occurs, it allows evaluation of the non-point source of pollution entering the receiving water. If needed, this source may be used in subsequent fate and transport modeling. Because numerous studies have identified sediments as potential reservoirs of microorganisms in stormwater, sediment samples, in addition to water samples, should be collected at the outfall. Sampling elsewhere in the receiving water is not recommended because the frequency of sampling is intense (as described below) and the number of samples and assays needed quickly grows to become infeasible. The receiving waterbody is either a river, lake or the coastal ocean where recreational water activities are known to occur. A stormwater conveyance that has been previously studied or characterized should be given priority.

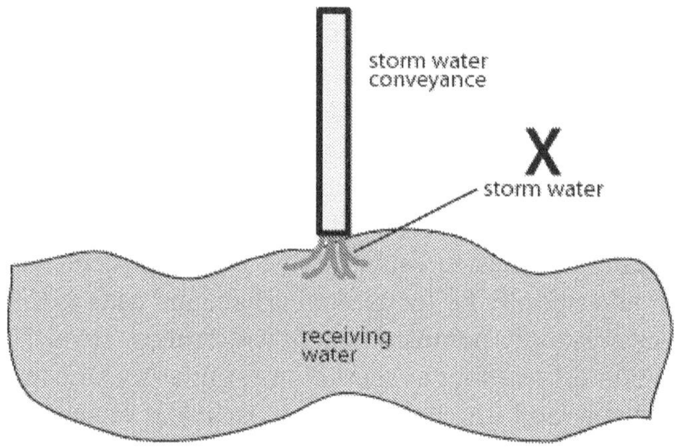

Figure 7-3. Schematic of Location to be Sampled at a Generic Sampling Site.

7.6.2.3 Sampling Type and Technique

Prior to conducting stormwater sampling, the appropriate sampling type (grab or composite) and technique (manual or automated) should be selected. Sampling type and technique are typically dictated by costs, maximum holding times for samples and project goals. For this study, it is recommended that standard stormwater outfall monitoring (U.S. EPA, 1992b) be used. Additionally, it is recommended automated samplers be used to collect stormwater sample, although consideration should be given to maximum holding times and potential cross contamination between storm events. Automated sampling may be more expensive, however it is often superior when conditions fluctuate rapidly or sporadically, or when available personnel are lacking. Many types of automatic samplers exist and none is ideal for all situations. Commonly used water samplers are described by Burton and Pitt (2001).

7.6.2.4 Inter-Storm and Intra-Storm Variation

Sample timing and duration should be dictated by previous work within the watershed on stormwater discharge. Scientists or managers should conduct a literature review and personal interviews with knowledgeable individuals to assess what is known about stormwater and stormwater pollutant variability among (inter-) and within (intra-) storms to guide each field study design. Caution should be taken when applying previous information regarding pollutant variability to microorganisms in stormwater. Microorganisms in stormwater vary in size and are not dissolved; hence, information about particulate pollutant variability might be extendable to microorganisms, while information about dissolved pollutant variability may not. In addition, microorganisms in stormwater may or may not be associated with suspended particles and this may impact variability.

Unless enough information is known about inter-storm variability in the chosen watershed to conduct a power analysis, we estimate that at least eight storms should be sampled over two rainy seasons. Storms should be selected based on the length of antecedent dry weather (ADW) and storm intensity. These two stormwater characteristics have been chosen to study inter-storm variability because they can influence the concentrations of other pollutants in stormwater (Burton and Pitt, 2001; Soller et al., 2004a). The identification of a particular type of storm (based on these two criteria) within a watershed that yields greater risks (due to having high concentrations of pathogens or greater exposure) could lead to valuable guidelines that relate risk to watershed storm characteristics, and also could be used to guide watershed prioritization for the national stormwater monitoring program. Table 7-8 shows the combinations of these characteristics and the number of each combination that should be targeted during the pilot study. These four types of storms are described as 1) low intensity, short period of ADW; 2) low intensity, long period of ADW; 3) high intensity, short period of ADW; and 4) high intensity, long period of ADW. Definitions of high and low rainfall intensity and long and short ADW period will need to be determined for each watershed investigated. Two storms are chosen for each combination so that inter-storm comparisons can be made statistically. In Table 7-9, showing different types of storms based on rainfall intensity and ADW period within a watershed, definitions of high and low rainfall intensity and long and short ADW period will depend on the particular watershed.

Table 7-9. Recommended Number of Storm Events that Should be Sampled for Each Type of Storms Based on Rainfall Intensity and ADW Period within a Watershed.

Storm intensity/ ADW duration	Low	High
Long	2	2
Short	2	2

For a particular storm, one may divide the resulting stormwater discharge into several types relevant to risk assessment: 1) initial discharge attributable to rainfall (beginning of storm), 2) highest concentration in stormwater during a rain event, and 3) discharge after a rain has ceased. These are all potential targets for sampling in a study aimed to quantify risk associated with exposure to stormwater. Waters during the beginning and after the rain event are important to sample because these represent times when exposure frequency is likely to be high (individuals are still in the water when the rain starts, or have returned to the water after the rain has ceased). It is important to characterize the highest concentration during a storm as this is when the potential dose for an exposure event is highest. Unfortunately, there is no way to predict, a priori, when the concentrations of pathogens will be the highest during a storm. Concentrations of contaminants in stormwater are non-steady and any patterns observed are likely to be site and storm specific. This implies that sampling will need to continue throughout the storm at regular intervals to understand the range of risks associated with exposure to stormwater (listed as Objective 1 under study aims – Table 7-1).

If possible, sampling frequency should be informed by previous work conducted in the watershed (Burton and Pitt, 2001). If no information is available, samples should be collected at the end of the stormwater conveyance every 15-30 minutes for the entire duration of the storm and for two days after the storm (Figure 7-2). Sampling after the storm is necessary, as exposure is likely to be relatively high, and storm water will continue to flow. It is understood that the sampling frequency may need to be scaled back given storm lengths and available resources. Sediment samples should be collected at the beginning, middle, and after the storm. Sediment sampling is not as intense as water sampling because less temporal variability is expected in sediment than in water samples. Given the vast literature on pathogens and indicators accumulating in sediment, it is important to sample sediment as it may serve as an "integrator" of stormwater quality or a source of pathogens itself.

Additionally, the storm sewer outfall or storm conveyance should be metered such that flow rate can be quantified and rainfall should be collected at the pilot sampling location site. An increase in flow rate over background base flow levels will signal the beginning of a storm event. Flow-paced sampling, often useful for determining the total mass flux of a contaminant into a receiving waterbody, is not recommended because instantaneous concentrations are more important than average concentrations. Therefore, time-paced sampling, which could be achieved with an automated sampler (for example, the ISCO automated sampler) is recommended, although maximum sample holding times must be considered.

7.6.2.5 Pathogen Enumeration

Methods developed as part of Phase 1b of the phased projects to enumerate important pathogens and indicator organisms in stormwater should be applied to the water and sediment samples collected during this phase of the pilot study. As described in Phase 1, it is imperative that the pathogen enumeration procedure contain a step to account for suspended-sediment sorbed pathogens. Each one of the microorganisms identified as target pathogens, as well as a suite of indicators (coliforms (including *E. coli*), *Enterococcus*, coliphage, and Bacteroides) should be assayed in all stormwater and sediment samples.

7.6.2.6 Ancillary Data

Instruments should be deployed to measure both rainfall and volumetric flow rate as functions of time in the storm sewer or stormwater conveyance. Grab samples, which will be used to assay microorganisms, should also be used to assay chemical and physical indicators of water quality including temperature, salinity, nitrate, ammonium, soluble reactive phosphate, turbidity, and total suspended solids. Sediment samples should be analyzed for clay, silt and sand content, as well as for organic carbon components.

7.6.2.7 Exposure Frequency

From the onset of the storm until sampling is complete 2-days after the end of the rain event, the number of individuals recreating in the receiving water and their activities should be recorded. If the area of exposure is small enough, this may be accomplished using a time-lapse video camera. The activity types and exposure frequencies can be used to 1) estimate the volume ingested, and 2) determine the population-based risk of illness using the dynamic risk assessment models.

7.6.2.8 Data Analysis

Using the measured concentrations of pathogen during the storm and the recorded exposure frequency, along with an assumed dilution and die-off factor, one can estimate the risk of illness from exposure to a particular pathogen during a storm event for individuals in a receiving water. Figure 7-4 (left) shows hypothetical time series of flow discharge, infectious enterovirus and viable *Enterococcus* concentrations, and individuals exposed to stormwater adjacent to a stormwater outfall, such as the one shown in Figure 7-3. These data should be used in conjunction with the risk equations presented in Chapter 8.0 to determine a time series of risks attributable to stormwater exposure. The range of risks could be compared to risk targets for recreating in freshwater and marine waters (8/1000 and 19/1000, respectively).

The data collected for a particular storm can help determine if there is a relationship between pathogens and fecal indicators. Figure 7-4 (below) shows a hypothetical plot of viable *Enterococcus* and infectious enterovirus; one could use such a plot to look for relationships between indicators and pathogens. In addition, Spearman rank correlation coefficients (non-parametric) can be used. Multiple indicators (coliforms, *Enterococcus*, phage, and/or Bacteroides) can also establish whether there is a correlation between increased levels of various combinations of indicators and pathogen concentrations.

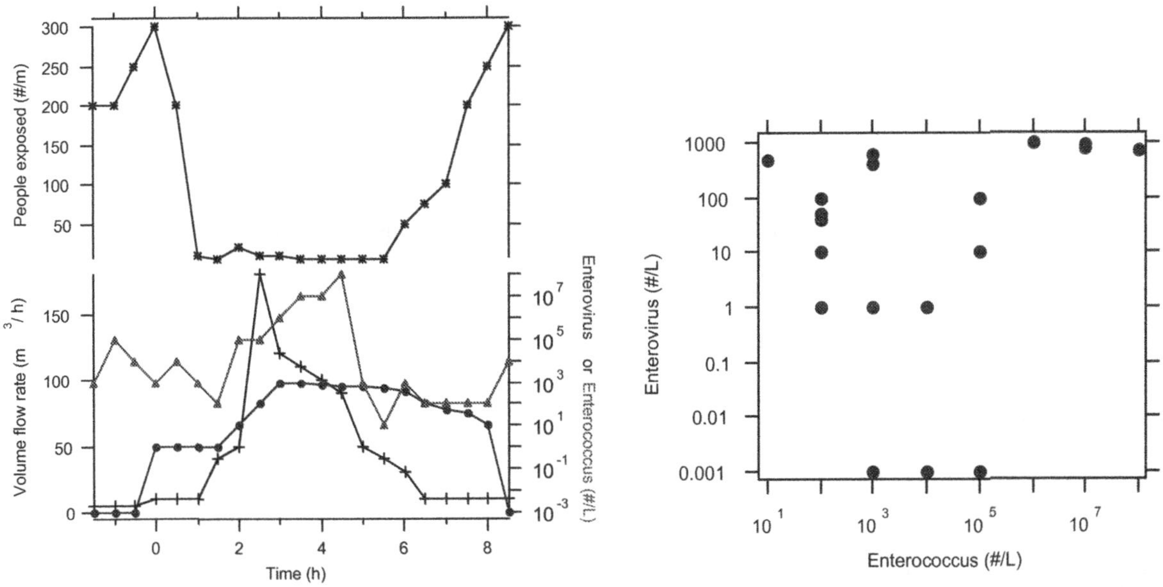

Figure 7-4. (Left) A Hypothetical Graph of Enterovirus Concentrations (•), *Enterococcus* Concentrations (∧) and Flow Rate (+) During a Storm. The number of individuals exposed to stormwater adjacent to the conveyance is shown as a function of time before, during and after the storm. At t=0, the rain begins, at t=5.5 h, rain stops. (Right) A plot of a pathogen (enterovirus) as a function of an indicator concentration (*Enterococcus*).

The risk distributions from the pilot study can be used to examine if and how storm characteristics influence median, average, and ranges or risks. Such relationships could subsequently be used to target specific storm types for further studies. These types of relationships could also be used to assign a "concern level" for storms, similar to that presented by Soller et al. (2006) for Newport Bay, California.

7.7 Phase 3: National Stormwater Monitoring Program and Risk Assessments

The overall goals of a national stormwater monitoring program will be to quantify: 1) the concentration of target infectious/viable pathogens in stormwater at the "end of the pipe", and 2) the number of individuals exposed to stormwater in receiving waters for various types or classes of stormwater. The previous section described a pilot-scale investigation for a single watershed that will yield data to help determine the total number of watersheds that need to be studied to conduct a national monitoring program useful for risk assessment purposes. The pilot study will also provide valuable insights regarding inter- and intra-storm variability of pathogens in stormwater and how this will lead to variability in terms of risk. These insights will aid in the design of monitoring programs for other watersheds, although it should be noted that results regarding inter- and intra-storm variability in the pilot watershed may not directly apply to other watersheds. The pilot scale study protocol recommends sampling over eight storms with samples collected every 15-30 min during the storms. One of the primary issues in the pilot scale study is whether the number of samples collected per storm, or the number of storms samples, can be reduced without losing valuable information. This question can only be addressed with the pilot-scale data in hand.

In deciding which watersheds to include in a national monitoring program, it may be important to consider all land uses, including agricultural and forested, as wild animal feces in these sub-watersheds areas may represent an important source of contamination. Table 7-10 shows various types of stormwater present in the U.S. It assumes that land use and climate are the main drivers for concentrations of microorganisms in stormwater. Each column represents a dominant land use; each row is a particular climate in the U.S. In parentheses is a suggested state where the directed study might take place.

Table 7-10. Recommended Sampling Locations for the National Stormwater Monitoring Program Based on Various Types of Stormwater Present in the U.S.

Climate (Location in the U.S.) /Dominant Land Use	Industrial	Residential	Commercial	Agricultural	Forested
Arid Southwest(California)		X	X	X	
Continental Climate of Midwest (Great Lake States)	X	X		X	
Continental Climate of Northeast (New Jersey)	X	X			X
Subtropical Southeast (Florida)		X	X		X

Testing all combinations of land use and climate (Table 7-10), would require studies at 25 sites. Given the extensive inter-storm and intra-storm comparisons for each watershed, it is likely that a subset of these 25 combinations would suffice for monitoring. In Table 7-10, 10 Xs appear in the boxes representing the combinations of land use and climate that the Project Team suggests should receive consideration for the national monitoring program. Objectives (2) and (4) presented in Table 7-1, will likely require at least two studies in each land use / climate category so that comparative analyses can be conducted using analyses of variance or chi-squared analyses.

It is not possible to characterize the full scope of a national monitoring program at this time. However, the issues described above, along with those considered in development of the pilot monitoring program, provide an outline, context, and information that will allow development of the national monitoring program. For example, based on the results of the pilot monitoring, it may be possible to conduct a power analysis to determine how many storm events will need to be monitored at each location, and how many different locations will be required given the inter- and intra-location variability observed during the pilot investigation.

Once the national stormwater monitoring program is completed, the next goal of this component of the recommended plan of action will be to develop a generalizable risk model that relates risk to easily quantifiable watershed characteristics. This will enable researchers to give high priority to specific watershed types for future study and monitoring. use of the data collected during previous phases in this manner could help direct management actions, inform the potential utility of future epidemiology studies, and inform regulatory limits for microorganisms in stormwater.

7.8 Summary

The proposed effort consists of a series of phased projects. Phase 1a is the development of interim guidance to facilitate risk management and resource allocation prioritization decisions; Phase 1b focuses method development for enumerating infectious/viable pathogens in stormwater. It is important to consider methods for enumerating pathogens associated with suspended particles in the methods development component.

Once both aspects of Phase 1 are complete, a pilot monitoring program should be undertaken. Phase 2 is a pilot program for the intensive monitoring of an urban watershed for pathogens and indicator organisms in stormwater at the "end of the pipe." The objectives of the pilot study are the same as those of the national monitoring (Table 7-1). The pilot study will serve as a proof of concept for the national study (Phase 3) and will serve to evaluate the proposed frequency of sampling to document inter- and intra-storm variability. Sampling should commence at the very onset of each monitored storm and continue for an extended period after the storm. Sediment samples should also be analyzed. In addition to pathogen enumeration, the number of individuals recreating in the water should also be recorded. Sampling should be conducted over eight storms at each sampling site, thus encompassing a range of rain events defined by their antecedent dry weather period and the rain intensity.

Phase 3 is a national stormwater monitoring program intended to acquire the data necessary to characterize the range of risks associated with exposure to microorganisms in stormwater. Upon completion of the data collection, a risk assessment model would be developed and implemented.

- The specific intended outcomes of the national monitoring protocol are:
- Determination of the range in concentrations of key pathogens at the "end-of-the-

pipe" of stormwater drains, watershed outlets, or conveyances adjacent to receiving waters.

- Relating the presence and concentrations of pathogens and fecal indicators in stormwater to stormwater types, defined by land use, climate, antecedent dry weather period, and rainfall intensity.
- Relating the types of exposures and number of individuals exposed to stormwater to stormwater types, defined by land use, climate, antecedent dry weather period, and rainfall intensity.
- Conducting risk assessments and relating risk to stormwater types, defined by land use, climate, antecedent dry weather period, and rainfall intensity.
- Identifying "high risk" stormwater types for further study, consistent monitoring, or remediation.

APPENDIX A

GENESIS OF CURRENT MICROBIAL GUIDANCE LEVELS FOR RECREATIONAL WATERS

In 1914, the United States Public Health Service (U.S. PHS) set a standard requiring that drinking waters not show evidence of coliform organisms (U.S. Treasury Department, 1914). Shortly after the development of those drinking water standards, the U.S. PHS pursued concurrent investigations of the role of recreational and shellfish waters in enteric disease transmission by compiling data on outbreaks (Frost, 1925; Stokes, 1927a, 1927b). Stokes (1927a, 1927b) reported that a 1921 epidemic of typhoid fever was unquestionably attributable to bathing in polluted waters (NRC, 2004).

Around 1950, several proposed guidelines for recreational waters were being developed throughout the U.S. (APHA, 1949; Cox, 1951; Scott, 1951; Streeter, 1951). By the mid-1950s, a variety of bacterial indicator standards had appeared in the U.S. In 1956, Garber 1956 reported that the City of Los Angeles was conducting a survey of standards for recreational waters in 13 jurisdictions. The State of Illinois used *Enterococcus*, while the remaining jurisdictions used total coliforms. Statistical reporting varied over a wide range: eight jurisdictions used either a geometric mean or a median for total coliforms; three used an arithmetic mean; four used a percentage that could not be exceeded; and three used absolute maximums. The most common standard was a requirement that total coliforms not exceed 1,000 per 100 mL (NRC 2004).

Of particular importance were the studies conducted in the late 1940s and early 1950s by the U.S. PHS at bathing beaches on Lake Michigan at Chicago, Illinois; on the Ohio River at Dayton, Kentucky; and on Long Island Sound at Mamaroneck and New Rochelle, New York (Stevenson, 1953). All of the studies followed a similar design. Two beaches with different water quality were selected at each location, except at the Dayton location where a beach with high quality water could not be found. At the Dayton location, a large public swimming pool was used as a substitute for the high quality beach. Each location was chosen because, in addition to beaches having suitable water quality, there was a large residential population nearby that used the beaches. Cooperating families used a calendar system that allowed then to record their swimming activity and illnesses on a daily basis for the entire summer. They recorded gastrointestinal, respiratory, and other symptoms such as skin irritations. The water quality was measured on a routine basis, using total coliform bacteria as the indicator organism (U.S. EPA, 1986).

The results of the Lake Michigan beach study indicated that there were no excess illnesses of any type in swimmers at beaches that had median coliform densities of 91 and 180 per 100 mL over a swimming season when compared to the number of illnesses in the total study population. The water quality similarity at the two Lake Michigan beaches was unexpected, as previous experience had indicated that there was a difference in water quality at the beaches. A second method of analysis compared the illness observed in the week following three days of high coliform density with that observed following swimming on three days of low coliform density. The analyses showed that there was a significantly greater illness rate in individuals who swam on the three days when the geometric mean coliform density was 2300/100 mL, as compared to illness rates in swimmers who swam on the three days when the geometric mean coliform density was 43 per 100 mL. A difference was not observed when the geometric mean coliform density on high and low days was 732 and 32 per 100 mL respectively. Data from the Ohio River study indicated that swimmers who swam in water with a median coliform density of 2300 coliform per 100 mL had an excess of gastrointestinal illness when compared to an expected rate calculated from the total study population. No other associations between swimming and illness were observed. The results of two marine bathing beach studies showed no association between illness and swimming in water containing 398 and 815 coliforms per 100 mL (U.S. EPA, 1986).

In 1968 the National Technical Advisory Committee (NTAC), convened by the U.S. Federal Water Pollution Control Administration was charged with proposing microbiological criteria for recreational waters. The NTAC determined that a fecal coliform measurement was more appropriate than a total coliform measurement because the fecal coliform measurement is more specific. The coliform water quality index used during the U.S. PHS epidemiological studies was translated into a fecal coliform index by using the ratio of fecal coliforms to coliforms (~18%) at the location on the Ohio River where the original study had been conducted in 1949. using this proportion, the NTAC found that the equivalent of 2300 coliforms per 100 mL, the density at which a statistically significant swimming-associated gastrointestinal illness was observed, was about 400 fecal coliforms per 100 mL. The NTAC suggested that a detectable risk was undesirable and, therefore proposed one-half of that density, 200 fecal coliform per 100 mL, as the density at which a health risk occurred. The NTAC also suggested that the use of the water should not cause a detectable health effect more than 10% of the time. Thus, they established the recommended criterion for recreational waters (log mean of 200/100mL and 90th percentile of 400/100mL) (U.S. EPA, 1986).

The U.S. EPA (1976) recommended this criterion again in 1976 even though it had been criticized on a number of issues. Henderson (1968) published one of the earliest critiques of the recommended criterion. He noted the paucity of epidemiological data in support of any numerical ceilings based on fecal indicators and criticized the one proposed as to the poor quality of the database, the derivation of the specific limits and the indicator system used. Moore (1975) objected to the selection of only part of the data from the Lake Michigan study to develop the 200 fecal coliforms per 100 mL recreational water criterion. He observed that the agency had ignored opposite findings in the Lake Michigan studies. He pointed out that the inclusion of all illnesses reported during the week after a bathing episode made the association of these ailments with the bathing episode tenuous. He also claimed there was no way of knowing how the incidence of skin irritations in bathers who swam on clean days compared to the frequency of diarrhea in those who swam on other days, because all the illnesses reported were lumped together (U.S. EPA, 1986).

Cabelli et al. (1985) suggested other weaknesses in the U.S. PHS study design which would have precluded the identification of swimming-associated, pollution-related illnesses if, in fact, they occurred. They pointed out that "swimming" was "poorly defined and that it was unknown whether “swimming” study participants actually immersed their bodies, much less their heads, in the water. They also criticized this limitation and the use of the calendar method for recording "swimming" episodes and illnesses as precluding the inclusion of beach-going but nonswimming control groups in the studies. Moreover, the use of the calendar approach with nearby residents and the day-to-day variability in the pollution levels at the beaches increased the probability of a given individual's exposure to different levels of pollution during the incubation period of the illness.

This criticism ultimately resulted in a National Research Council (NRC) opinion that "No specific recommendation is made concerning the presence or concentrations of microorganisms in bathing water because of the paucity of valid epidemiological data" (NRC, 1972). The NRC also criticized the fecal coliform measurement itself because thermotolerant bacteria, such as Klebsiella spp. are included in this group but are not necessarily fecal in origin.

The U.S. EPA, in 1972, initiated a series of studies at marine and fresh water bathing beaches, designed to correct the perceived deficiencies of the U.S. PHS studies. One goal of the U.S. EPA studies was to determine if swimming in sewage-contaminated water carries a health risk for bathers; and, if so, to what type of illness. If a quantitative relationship between water quality and health risk was obtained, two additional goals were to determine which bacterial indicator is best correlated to swimming-associated health effects and if the relationship is strong enough to provide a criterion (U.S. EPA, 1986).

The marine studies were conducted at bathing beaches in New York City, New York, Boston, Massachusetts, and at Lake Pontchartrain, near New Orleans, Louisiana. Two beaches were selected at each site; one that received very little or no contamination and the other whose water quality was barely acceptable with respect to local recreational water quality standards. In the New York City and Boston Harbor studies, the "barely acceptable" beaches were contaminated with pollution from multiple sources, usually treated effluents that had been disinfected (U.S. EPA, 1986).

The freshwater studies were conducted on Lake Erie at Erie, Pennsylvania and on Keystone Lake outside of Tulsa, Oklahoma. The "barely acceptable" beaches at both sites were contaminated by effluents discharged from single point-sources (U.S. EPA 1986).

In these studies, subjects self-diagnosed all reported symptoms. The results of these studies indicated that symptom categories unrelated to gastroenteritis usually did not show a significant excess of illnesses at either of the paired beaches at each study location. Moreover, significantly increased swimming-associated gastroenteritis rates were always observed at the more polluted of the paired beaches at each study location. Statistically significant swimming-associated gastroenteritis rates were not observed at any of the relatively unpolluted beaches. The occurrence of a statistically significant excess of swimming-associated gastroenteritis in swimmers who bathed at beaches that were more polluted is indicative that there is an increased risk of illness from swimming in water contaminated with treated sewage (U.S. EPA, 1986). This finding, which was observed at both marine and fresh water locations was important because it placed in proper perspective the relationship between water contaminated with treated sewage and health risks for swimmers. This association was not very well defined in earlier studies (U.S. EPA, 1986).

The data from the three years of the New York City study (marine water) were analyzed in two ways. The first was by grouping trial days with similar indicator densities from a given swimming season and the second was by looking at each entire summer. For both analyses, *Enterococcus* showed the strongest relationship to gastroenteritis. *E. coli* was a very poor second and all of the other indicators, including total coliforms and fecal coliforms showed very weak correlations to gastroenteritis. Subsequent studies, including the freshwater trials, used

Enterococcus and *E. coli*. The researchers also included fecal coliforms in subsequent studies because of their status as an accepted criterion (U.S. EPA, 1986).

The freshwater studies were analyzed only by summer. The correlation coefficient for *E. coli* was slightly greater than that for *Enterococcus*, however, statistical analysis indicated that the two values were not significantly different. Fecal coliforms showed no correlation to swimming associated gastroenteritis rates. The similarities in the relationships of *E. coli* and *Enterococcus* to swimming associated gastroenteritis in freshwater indicate that these two indicators are equally efficient for monitoring water quality in freshwater, whereas in marine water environments only *Enterococcus* provided a good correlation (U.S. EPA, 1986).

The quantitative relationships between the rates of swimming associated health effects and bacterial indicator densities were determined using regression analysis. Linear relationships were estimated from data grouped by summers or trials with similar indicator densities. The data for each summer were analyzed by pairing the geometric mean indicator density for a summer bathing season at each beach with the corresponding swimming-associated gastrointestinal illness rate for the same summer. The swimming-associated illness rate was determined by subtracting the gastrointestinal illness rate in nonswimmers from that for swimmers. These two variables from multiple beach sites were used to calculate a regression coefficient, y-intercept and 95% confidence intervals for the paired data. In the marine studies the total number of points for use in regression analysis was increased by collecting trial days with similar indicator densities from each study location and placing them into groups. The swimming-associated illness rate was determined as before, by subtracting the nonswimmer illness rate of all the individuals included in the grouped trial days from the swimmer illness rate during these same grouped trial days (U.S. EPA, 1986).

U.S. EPA did not recommend a change in the stringency of its bacterial criteria for recreational waters (U.S. EPA, 1986). EPA's evaluation of the bacteriological data indicated that using the fecal coliform indicator group at the maximum geometric mean of 200 per 100 mL would cause an estimated 8 illnesses per 1,000 swimmers at fresh water beaches and 19 illnesses per 1,000 swimmers at marine beaches. EPA indicated that these relationships are only approximate and are based on applying ratios of the geometric means of the various indicators from the U.S. EPA studies to the 200 per 100 mL fecal coliform criterion, and that these are U.S. EPA's best estimates of the accepted illness rates for areas which apply the U.S. EPA fecal coliform criterion (U.S. EPA, 1986).

The current *E. coli* and *Enterococcus* criteria (U.S. EPA, 1986) were developed as log-linear relations between indicator densities and the previously noted levels of accepted illness rates. The equations developed by Dufour and Cabelli were used to calculate the geometric mean indicator densities corresponding to the accepted gastrointestinal illness rates for steady state dry weather conditions (U.S. EPA 1986). Single-sample maximum allowable densities were also promulgated relating allowable single sample maximum values to the relative level of use expected to occur at recreational sites.

APPENDIX B

EXAMPLE OF A WEB-BASED QUESTIONNAIRE, TRANSMITTAL FORM AND RESULTS

~ REQUEST FOR STORMWATER MICROORGANISM DATA ~

Currently, it is not uncommon for stormwater NPDES permits to include monitoring of bacterial indicator organisms in receiving waters as means of identifying impacts on beneficial uses and developing best management practices to minimize the risk to human health. However, the relationship between concentrations of indicator organisms and pathogens in stormwater is not well characterized. Furthermore, the basis for most state and federal guidance levels of indicator organisms in water (including stormwater) is from studies which investigated health effects associated with full body contact recreation in surface waters impacted by either wastewater effluent. Thus, the applicability of the existing ambient water quality criteria warrants critical consideration with respect to stormwater.

To assist in developing guidance that will enable more accurate and defensible evaluations of stormwater microorganism data and the associated risk to human health from exposure to pathogenic microorganisms in stormwater, WERF is currently conducting an investigation that entails gathering relevant stormwater pathogen and indicator information. Specifically, WERF is requesting pathogen and indicator data relevant to separate stormwater systems and/or receiving waterbodies that your agency or organization has collected and analyzed.

The types of data we are primarily interested in are waterborne pathogen and/or fecal indicator data collected in separate stormwater systems, including stormwater outfalls, urban runoff, and receiving waterbodies. If your agency has information that you believe may be useful to the investigation, please complete the data collection form located at http://www.eoainc-w2k.com/werf_survey/werf-survey.asp.

Feel free to contact me with any questions, comments, or concerns via telephone, fax, or e-mail. Thank you in advance for your assistance in this important investigation.

Sincerely,

Project Manager
Organization
Phone Number
E-mail Address

Name of Organization or Agency ________________________

Type of Agency

- ❑ Public Utility/District
- ❑ City/County
- ❑ Research Organization/University
- ❑ Other: ________________________

Available Data

Pathogens

- ❑ *Salmonella nontyphodial*
- ❑ *Shigella* spp.
- ❑ *Campylobacter* spp.
- ❑ *Cryptosporidium parvum*
- ❑ *Giardia lamblia*
- ❑ *Toxoplasma gondii*
- ❑ Enterovirus
- ❑ Adenovirus
- ❑ HAV
- ❑ Norwalk Virus
- ❑ Astrovirus
- ❑ *Pseudomonas aeruginosa*
- ❑ Others________________________
  ____

Indicator Organisms

- ❑ Fecal Coliform
- ❑ Total Coliform
- ❑ *Enterococcus*
- ❑ *E. coli*
- ❑ Others________________________

Contact Person

Name: ________________________
Address: ________________________
____
____
Phone #: ________________________
Fax #: ________________________
e-mail: ________________________

Location of Sampling	❑ Receiving Waterbody (e.g., Creek) ❑ Stormwater Outfall ❑ Stormwater Runoff ❑ Other: ____________________
Frequency of Sampling	❑ Daily ❑ Weekly ❑ Monthly ❑ Quarterly ❑ Other: ____________________
Number of Sampling Locations	____________________
Sampling Timeframe	____________________
Was QA/QC conducted?	❑ Yes ❑ No
Were data published?	❑ Yes ❑ No
If yes, please provide citation	____________________
Current format of data	❑ Excel/Database ❑ Hard copy ❑ Other: ____________________
Please list any additional information or comments	____________________ ____________________ ____________________ ____________________ ____________________ ____________________

	Agency/Org Name	Pathogens						Indicator Microorganisms					Sample Type			
		Samonella	Campylobacter	Cryptosporidium	Giardia	Enterovirus	Adenovirus	Fecal Coliform	Total Coliform	Entero-coccus	E. Coli	Other Indicator	Receiving Waters	Stormwater Outfalls	Stormwater Runoff	Other
1	Philadelphia Water Department			x	x			x	x	x	x		x	x	x	
2	City of Encinitas						x	x	x	x			x	x	x	
3	University of Auckland	x	x			x	x	x	x	x	x	x	x	x		
4	The City of Edmonton			x	x			x	x		x		x	x		
5	LAS VIRGENES MUNICIPAL WATER DISTRICT			x	x	x		x	x				x			
6	U.S. Geological Survey			x									x			x
7	Green Bay Metro Sewerge District			x	x			x	x	x	x	x	x			
8	City of Tulsa							x					x	x	x	
9	City of Santa Cruz, Public Works Department							x	x	x			x	x	x	
10	Port of Seattle							x		x	x	x	x	x	x	
11	Capacity (Wellington Water Management)							x		x	x		x	x	x	
12	Virginia Tech							x	x	x	x		x	x	x	
13	Municipality of Anchorage, Watershed Management Services							x	x			x	x	x	x	x
14	Southern California Coastal Water Research Project								x	x	x		x	x	x	
15	University of South Florida							x		x	x	x	x	x	x	
16	City of Santa Barbara								x	x	x		x	x	x	
17	Wichita Water & Sewer Department							x			x		x	x	x	
18	Institut fuer Seenforschung							x	x	x	x	x	x	x	x	x
19	City of Redmond							x			x		x	x		
20	Storm Water Management Authority, Inc							x					x	x		
21	Anacostia Watershed Society							x			x		x	x		
22	County of Orange, Watershed and Coastal Resources Division							x	x	x			x	x		
23	Metro Nashville Water Services										x		x	x		
24	Urban Drainage and Flood Control District							x			x	x	x	x		x

	Agency/Org Name	Pathogens						Indicator Microorganisms					Sample Type			
		Samonella	Campylobacter	Cryptosporidium	Giardia	Enterovirus	Adenovirus	Fecal Coliform	Total Coliform	Entero-coccus	E. Coli	Other Indicator	Receiving Waters	Stormwater Outfalls	Stormwater Runoff	Other
25	Riverside County Flood Control & Water Conservation District							x	x	x	x		x	x		
26	Northeast Ohio Regional Sewer District							x	x		x	x	x	x		
27	City of Akron - Public Utilities Bureau							x			x		x	x		
28	Hampton Roads Sanitation District									x		x	x	x		
29	Upper Occoquan Sewage Authority							x						x		
30	City of Denton Water Utilities Department -Watershed Protection Div.							x						x		
31	Charleston Commissioners of Public Works							x		x		x		x		
32	City of Boise							x			x			x		
33	City of Scottsdale							x				x		x		
34	Orange County							x	x	x			x		x	
35	City of Santa Barbara Creeks Restoration/Water Quality Improvement								x	x	x		x		x	
36	Pinellas County Department of Environmental Management							x	x	x			x		x	
37	BOSTON WATER AND SEWER COMMISSION							x		x	x		x		x	x
38	Stormwater Management Inc.							x	x						x	
39	City of Indianapolis, Department of Public Works, Office of Environmental							x			x		x			x
40	Macon Water Authority							x					x			
41	3 Rivers 2nd Nature Project, STUDIO for Creative Inquiry, Carnegie Mellon							x	x	x	x		x			
42	Johnson County Wastewater/Stormwater							x		x	x		x			
43	CITY OF OXNARD								x	x	x		x			
44	U.S. Geological Survey							x					x			
45	The Scotts Miracle-Gro Company, Hyponex Corporation								x		x		x			

	Agency/Org Name	Pathogens						Indicator Microorganisms					Sample Type			
		Samonella	Campylobacter	Cryptosporidium	Giardia	Enterovirus	Adenovirus	Fecal Coliform	Total Coliform	Entero-coccus	E. Coli	Other Indicator	Receiving Waters	Stormwater Outfalls	Stormwater Runoff	Other
48	City of Port Huron Water Reclamation Facility								x		x		x			
49	Littleton/Englewood WWTP							x			x		x			
50	City of Boise							x			x		x			
51	Clackamas County Water Environment Services										x		x			
52	Fairfax County (VA) Dept. of Public Works and Environmental Services-							x		x	x		x			
53	Louisville and Jefferson County (Kentucky) Metropolitan Sewer District							x					x			x
54	Sonoma County Water Agency							x				x	x			
55	Metropolitan Sewer District of Greater Cincinnati							x			x		x			
56	Sanitation Districts of Los Angeles County															
57	Metropolitan Council Environmental Services							x								x
58	New York City Department of Environmental Protection							x	x	x						x
59	New York City Department of Environmental Protection							x								x
60	University of Manitoba							x	x							x

APPENDIX C

RESULTS OF LITERATURE REVIEW ON THE CONCENTRATIONS OF FECAL INDICATORS IN STORMATER

Indicator	Range	Location	Source	Waterbody Type	Detection Method	Storm Event	CSO/SSO	Notes
Total Coliforms	4-25,000 CFU/100mL	*Southern CA*	*Jiang and Chu, 2004*	*urban rivers*	standard membrane filtration method	wet season, 6 storm events detected	ND	
	1,000-1,000,000 MPN/100mL	San Diego, CA	Schiff and Kinney, 2001	storm drain	membrane filtration	storm events	ND	
	1,110-88,000 CFU/100mL	Germany	Kistemann et al., 2002	watercourse and tributary	CFU-membrane filtration	at the onset of heavy rainfall and runoff	one of three sites are possibly influenced by CSO	median
	204-3,156 CFU/100mL	Oak Creek, Arizona	Rose et al., 1987	creek	membrane filtration	rainy season	ND	average
	4,000-27,000 CFU/100mL	Orange County, California	Dwight et al., 2002	coastal waters	standard methods	rainy season	ND	
	630,000 CFU / 100 mL	Southern California	Noble et al., 2004	coastline	membrane filtration	1 day after storms of 2.7 to 7.6 cm of rain	ND	maximum
	1,400-1,450 CFU/100mL	California & Mexico	Noble et al., 2003	diverse waterbodies	FIB detection methods varies among different collaborating labs, including membrane filtration, multiple tube fermentation, and the defined substrate technology test kits, Colilert and Entercolert.	following a storm	ND	
	160,000-2,400,000 CFU/100mL	San Diego, CA	Schiff et al., 2002	creek	standard methods	3 storms sampled	ND	mean
	3.7×10^4 CFU/100mL	Monmouth County, NJ	Borst and Selvakumar, 2003	stormwater	Culture - CFU	1.8mm precipitation over 74 minutes	ND	
	960 CFU/100mL	Trenton, N.J.	Atherholt et al., 1998	river	Culture - CFU	precipitation throughout the year	ND	geomean for all year
Fecal Coliforms	188-356 CFU/100mL	North Carolina	Mallin et al., 1999	estuary creeks	membrane filtration method-CFU (mFC)	NA (monthly sample)	ND	geomean
	<1-4,000 CFU/100mL	Southern CA	Jiang and Chu, 2004	urban rivers	standard membrane filtration method	wet season, 6 storm events detected	ND	
	2,000-60,000 MPN/100mL	Louisiana	Jin et al., 2003	drainage canals, sample at storm	multiple tube fermentation technique	4 storm events	ND	Min/Max

Indicator	Range	Location	Source	Waterbody Type	Detection Method	Storm Event	CSO/SSO	Notes
				water pump station				
	102-105 MPN/100mL	San Diego, CA	Schiff and Kinney, 2001	storm drain	membrane filtration	storm events	ND	
	5,000-28,000 MPN/100mL	Louisiana	Jin et al., 2004	drainage canals, sample at storm water pump station	multiple tube fermentation technique	5 storm events	ND	Median
	1,000-10,000 CFU/100mL	A coastal Mediterranean river	Baudart et al., 2000	coastal rivers	membrane filtration	multiple storm events	CSO	
	1.7-7,000 CFU/100mL	West Virginia	Boyer and Kuczynska, 2003	spring water	membrane filtration	sampled monthly and seasonally	ND	
	96-470 CFU/100mL	Oak Creek, Arizona	Rose et al., 1987	creek	membrane filtration	rainy season	ND	average
	6,700-270,000 CFU / 100 mL	Israel	Asaf et al., 2004	rainfall & stormwater drains	culture - CFU	46 rain events in 2000-2003	ND	
	9-3,000,000 CFU / 100 mL	Sydney	Butler and Ferson, 1997	stormwater outlets	culture - CFU	sampled during storm events	ND	
	0-100 CFU/100mL	Southern Italy	Celico et al., 2004	springwater	culture - CFU	wet season	ND	
	1,315-6,400 CFU/100mL	Oak Creek, AZ	Crabill et al., 1999	Creek	culture - CFU	26mm rainfall the previous night	ND	maximum
	50,000 CFU / 100 mL	Southern California	Noble et al., 2004	coastline	membrane filtration	1 day after storms of 2.7 to 7.6 cm of rain	ND	maximum
	80-515 CFU/100mL	CA & Mexico	Noble et al., 2003	diverse waterbodies	membrane filtration, multiple tube fermentation, and the defined substrate technology test kits, Colilert and Entercolert.	following a storm	ND	
	14,000-16,000 CFU/100mL	San Diego, CA	Schiff et al., 2003	creek	standard methods	3 storms sampled	ND	mean
	$1.3x10^4$ / 100 mL	Monmouth County, NJ	Borst and Selvakumar, 2003	stormwater	culture - CFU	1.8mm precipitation over 74 minutes	ND	

Indicator	Range	Location	Source	Waterbody Type	Detection Method	Storm Event	CSO/SSO	Notes
	350-11,000/100mL	Canadian Stormwater, from various literature sources.	Marsalek and Rochfort, 2004	NA	NA	NA	ND	Mean
	316,0000 - 7,900,000 /100mL	summary, from various literature sources.	Marsalek and Rochfort, 2004	NA	NA	NA	CSO	
	6,100-127,000 CFU /100mL	Pittsburgh, Pennsylvania	Gibson III et al., 1998	urban stream	culture	storm events	CSO	averages
	150 CFU/100mL	Trenton, NJ	Atherholt et al., 1998	river	culture - CFU	precipitation throughout the year	ND	geomean for all year
E. coli	200-500 CFU/100mL	North Carolina	Mallin et al., 1999	estuary creeks	CFU (mTEC)	NA (monthly sample)	ND	geomean
	1,450-40,000 MPN/100mL	Louisiana	Jin et al., 2003	drainage canals, sample at storm water pump station	IDEXX Quanti-Tray method	4 storm events	ND	Min/Max
	2,005-20,980 MPN/100mL	Louisiana	Jin et al., 2004	drainage canals, sample at storm water pump station	IDEXX Quanti-Tray method	5 storm events	ND	Median
	22-13,700 CFU/100mL	Germany	Kistemann et al., 2002	watercourse and tributary	CFU-membrane filtration	at the onset of heavy rainfall and runoff	one of three sites are possibly influenced by CSO	median
	6300 CFU / 100 mL	southern California	Noble et al., 2004	coastline	membrane filtration	1 day after storms of 2.7 to 7.6 cm of rain	ND	
	600-5,000 MPN/100mL	Ft. Lauderdale, FL	Solo-Gabriele et al., 2000	River	substrate technology Colilert-18	different antecedent rain conditions	ND	
	1.7×10^4 / 100 mL	Monmouth County, NJ	Borst and Selvakumar, 2003	Stormwater	Culture - CFU	1.8mm precipitation over 74 minutes	ND	
	10^3-5×10^5/100mL	Canadian Stormwater, from various literature sources.	Marsalek and Rochfort, 2004	NA	NA	NA	ND	Mean
	944,000 - 1,900,000 /100mL	summary, from various literature sources	Marsalek and Rochfort, 2004	NA	NA	NA	CSO	
	110 CFU/100mL	Trenton, NJ	Atherholt et al., 1998	river	Culture - CFU	precipitation throughout the year	ND	geomean for all year
Fecal Streptococcus	7.5-13,600 CFU/100mL	Germany	Kistemann et al., 2002	watercourse and tributary	CFU-membrane filtration	at the onset of heavy rainfall and runoff	one of three sites are possibly	median

WERF

Indicator	Range	Location	Source	Waterbody Type	Detection Method	Storm Event	CSO/SSO	Notes
							influenced by CSO	
	44-500 CFU/100mL	Oak Creek, Arizona	Rose et al., 1987	creek	membrane filtration	rainy season	ND	average
	240-300,000 CFU/100mL	San Diego, CA	Schiff et al., 2003	creek	standard methods	3 storms sampled	ND	mean
	$1.0x10^3$ CFU / 100 mL	Monmouth County, NJ	Borst and Selvakumar, 2003	stormwater	Culture - CFU	1.8mm precipitation over 74 minutes	ND	
Enterococcus	<1-10,000 CFU/100mL	Southern CA	Jiang and Chu, 2004	urban rivers	standard membrane filtration method	wet season, 6 storm events sampled	ND	
	2,000-80,000 MPN/100mL	Louisiana	Jin et al., 2003	drainage canals, sample at storm water pump station	IDEXX Quanti-Tray method	4 storm events	ND	Min/Max
	102~105 MPN/100mL	San Diego, CA	Schiff and Kinney, 2001	storm drain	membrane filtration	storm events	ND	
	2,419-34,410 MPN/100mL	Louisiana	Jin et al., 2004	drainage canals, sample at storm water pump station	IDEXX Quanti-Tray method	5 storm events	ND	Median
	0-94 CFU/100mL	Southern Italy	Celico et al., 2004	spring water	Culture - CFU	Wet season	ND	
	6.1×10^3 CFU / 100 mL	Australia	Davies and Bavor, 2000	wetland	Wetland	Culture - CPU	precipitations of 0 to 28.5 mm	SSO
	228-310 CFU/100mL	CA & Mexico	Noble et al., 2003	diverse waterbodies	FIB detection methods varies among different collaborating labs, including membrane filtration, multiple tube fermentation, and the defined substrate technology test kits, Colilert and Entercolert.	following a storm	ND	
Coliphage	<2-7597 PFU/100mL	Southern California	Jiang and Chu, 2004	urban rivers	*E. coli* ATTCC 15597 host, Plaque Forming Unit	wet season	ND	
	0.5-333.2 PFU/ 100 mL	Southern California	Jiang et al., 2001	urban rivers	*E. coli* ATTCC 15597 and HS (pFamp)R hosts, Plaque Forming Unit	wet season	ND	

Indicator	Range	Location	Source	Waterbody Type	Detection Method	Storm Event	CSO/SSO	Notes
F+Coliphage	<2-853 PFU/100mL	Southern California	Jiang and Chu, 2004	urban rivers	*E. coli* HS(pFamp)R host, Plaque Forming Unit	wet season	ND	
	<2-300 PFU/100mL	Southern California	Jiang et al., 2001	urban rivers	*E. coli* HS(pFamp)R host, Plaque Forming Unit	wet season		
Somatic coliphages	0-6310 PFU / 100 mL	Australia	Davies et al., 2003	stormwater	Culture - PFU	Wet season	SSO	
f-RNA coliphages	0-63 PFU / 100 mL	Australia	Davies et al., 2003	stormwater	Culture - PFU	Wet season	SSO	*Salmonella* typhimurium WG49 as a host

Pathogen	Range	Location	Source	Waterbody Type	Detection Method	Storm Event	CSO/SSO	Notes
BACTERIA								
Salmonella spp	0.57-3MPN/L	Baltimore, MD	O'Shea and Field, 1991	separate storm sewer	MPN	storm water	SSO	
	0 - 100 MPN/L	Mediterranean coast	Baudart and et al., 2000	river	filter and SS agar culture-MPN, confirmed with biochemical tests	has year long records, values corresponding to storm events	CSO	
	0.6-424.5 MPN/L	France	Lemarchand and Lebaron, 2002	river	MPN	NA	SSO	median
	1st site, 4/12 positive (5min interval); 2nd site, 1/12 positive	Burlington, Ontario	Qureshi and Dutka, 1979	creek	filter and SS agar culture, confirmed with biochemical tests	2 storm events monitored	CSO	
Pseudomonas aeruginosa	500-54000/ L	Toronto	O'Shea and Field, 1991	storm water catchments	NA	during storm	NA	median
	59-200 / L	Baltimore	O'Shea and Field, 1991	separate storm sewer	MPN	storm water	SSO	
	<10-100,000/ L	Burlington, Ontario	Qureshi and Dutka, 1979	creek	membrane filtration-CFU	12 storm events monitored	CSO	

Pathogen	Range	Location	Source	Waterbody Type	Detection Method	Storm Event	CSO/SSO	Notes
	3,400-12,000 CFU/L	Navesink River, NJ	Selvakumar and Borst, 2006	Storm drains	membrane filtration-CFU	14 storm events monitored over 2 yr; 1.8 –51.8 mm, 0.9-15 mm/hr	nd	
Staphylococcus aureus	120-1200 MPN / L		O'Shea and Field, 1991	separate storm sewer	MPN	storm water	SSO	
	46,000-180,000 CFU/ L	Navesink River, NJ	Selvakumar and Borst, 2006	Storm drains	membrane filtration-CFU	14 storm events monitored over 2 yr; 1.8 –51.8 mm, 0.9-15 mm/hr	SSO	
Clostridium perfringens	15-4300 CFU/L	Germany	Kistemann et al. 2002	watercourse and tributary	CFU-membrane filtration	at the onset of heavy rainfall and runoff	one of three sites are possibly influenced by CSO	median
	400 CFU / L	Australia	Davies and Bavor, 2000	wetland	membrane filtration using CP agar base supplemented with tryptose sulphite cycloserine	rain season, total daily rainfall for each 24-h period preceding sample collection ranged from 0 to 28.5mm	nd	
	73-3610 CFU/L	Oahu, HI	Fujioka, 2001	a channel and a stream	described by Bisson and Cabelli (1979)-CFU	Sampled at June and July from mountainous rain-fed streams	nd	averages
	270 CFU/L	Trenton, NJ	Atherholt et al., 1998	river	described by Armon and Payment (1987)	60 seasonal sample sets; rainfall events included	nd	geomean for all year
***Bacillus* spp. spores**	88.5-1570 CFU/L	Oahu, HI	Fujioka, 2001	a channel and a stream	membrane filtration-CFU	Sampled at June and July from mountainous rain-fed streams	nd	averages
VIRUSES								
Adenovirus	$8.8x10^2$ - $7.5x10^3$ genomes/L	Southern CA	Jiang et al, 2001	Rivers	PCR	Wet Season	nd	
	Presence in 8 of 12 samples	Southern CA	Jiang and Chu, 2004	Rivers	PCR	During storms	nd	
Hepatitis A	Presence in 10 of 12 samples	Southern CA	Jiang and Chu, 2004	Rivers	PCR	During storms	nd	
Enterovirus	0.69 - 19 PFU/L	N/A	O'Shea and Field, 1991	Sewer/storm water	Culture - PFU	N/A	CSO	
	0.28 MPN/L	Hanauma Bay, HI	Reynolds et al, 1998	Bay	MPN PCR	Wet Season	nd	
	0 – 0.075 MPN /L	Oak Creek, AZ	Rose et al, 1987	River	MPN/PFU	Wet Season	nd	
	0.01 – 0.02 MPN/L	Charlotte Harbor, NC	Lipp et al. 2001	Estuary	Tissue Culture	Wet Season	Septic Systems	
	Presence in 9 of 12 samples	Southern CA	Jiang and Chu, 2004	Rivers	PCR	During Storms	nd	

Pathogen	Range	Location	Source	Waterbody Type	Detection Method	Storm Event	CSO/SSO	Notes
Rotavirus	0-0.24 IFF/ L	Oak Creek, AZ	Rose et al, 1987	River	Immuno-Fluo. Foci	Wet Season	nd	
Protozoans								
Giardia	0-12.4 cysts/100L	Germany	Kistemann et al., 2002	tributaries	filtration, density flotation, and immunofluorescence	9 extreme rainfall/runoff events sampled, 12 h avg duration	nd	median
	Positive Result	Charlotte Harbor, NC	Lipp et al. 2001	estuary	immunofluorescence	Positive results during storm events	Septic systems	
	67-7,500 cysts/100L	Pittsburgh, PA	Gibson III et al., 1998	urban stream	immunofluorescence	1 storm sampled, 4h duration	CSO	averages
	50 cysts/ 100 L	Trenton, NJ	Atherholt et al. 1998	plant main pipe from river	immunofluorescence	sampled monthly and seasonally	nd	geometric mean
Cryptosporidium	1.30-17.05 oocysts/100L	Germany	Kistemann et al., 2002	tributaries	filtration, density flotation, and immunofluorescence	9 extreme rainfall/runoff events sampled, 12h avg duration	nd	median
	Positive Result	Charlotte Harbor, NC	Lipp et al. 2001	estuary	immunofluorescence	Positive results during storm events	Septic systems	
	11250-946250 oocyst/100 L	Madera County, CA	Tate et al. 2000	rangeland watershed	immunofluorescence	4 storms events	nd	1.0 m from fecal pats
	350-15680 oocysts/100 L	West Virginia	Boyer and Kuczynska, 2003	spring	immunofluorescence	sampled monthly and seasonally	nd	
	39-1,667 oocysts/ 100L	Pittsburgh, PA	Gibson III et al., 1998	urban stream	immunofluorescence	1 storm sampled, 4h duration	CSO	averages
	4-730 oocysts/100 L	France	Lemarchand and Lebaron, 2002	coastal river	immunofluorescence	wet season sampled	nd	median
	10 oocysts/100L	Trenton, NJ	Atherholt et al. 1998	plant main pipe from river	microscopy	sampled monthly and seasonally	nd	geometric mean

APPENDIX D

Brief Description of UC Davis Methods for PCR-Based Detection of Pathogens in Stormwater

This Appendix briefly describes the qPCR protocol described by Rajal et al. (2005). 100 L of water should be collected at each sample time. The water should be seeded with known quantities of either PP7 (a *Pseudomonas* phage) or similar icosohedral ssRNA phage (MS2 is not recommended because it may be present in stormwater already whereas PP7 likely will not) to assess viral recovery. As suggested by Rajal et al., seeding with this phage will allow calculation of virus recoveries at various points in the concentration process. An aliquot of sample (1 L) should be taken for culture-based assays for indicator organisms. Bacteria and protozoa recoveries can be determined by comparing culture-based results with qPCR results for indicator organisms. The remaining volume (approx. 99 L) should be subsequently sieved through a 38 μm sieve and then filtered using a tangential membrane filtration device with a MW cutoff of 50 KDa. Retentate should be retained and returned to the lab on ice, where it should be sonicated or blended to account for particle-associated microbes. Further concentration may be necessary (Rajal et al. 2005). The described concentration should result in a 1000X concentration. The sample should be split into at least three parts. One part will be for cryogenic storage, one part for RNA extraction and the other for DNA extraction. Magnetic bead nucleic-acid clean-up may be performed if assay exists per Thompson et al. (in press). RNA and DNA may be stored for further analysis.

REFERENCES

Abu-Lail, N.I., Camesano, T.A. Polysaccharide properties probed with atomic force microscopy. *Journal of Microscopy* 2003, *212* (3), 217-238.

Akin, E.W. Presented at the U.S. EPA Symposium on Microbial Health Considerations of Soil Disposal of Domestic Wastewaters. 1981.

Akin, E.W., Hill, W.F., Cline, G.B., Benton, W.H. The loss of poliovirus 1 infectivity in marine waters. *Water Res.* 1976, *10*, 59-63.

Albert, J.M., Munakata-Marr, J., Tenorio, L., Seigrist, R.L. Statistical evaluation of bacterial source tracking data obtained by Rep-PCR DNA fingerprinting of *Escherichia coli*. *Environ. Sci. Tech.* 2003, *37*, 4554-4560.

Alexander, L.M., Heaven, A., Tennant, A., Morris, R. Symptomatology of children in contact with sea water contaminated with sewage. *J. Epidem. Comm. Health.* 1992, *46(4)*, 340-4.

Alm, E.W., Burke, J., Spain, A. Fecal indicator bacteria are abundant in wet sand at freshwater beaches. *Water Research* 2003, *37*, 3978-3982.

American Public Health Association (APHA). Standard Methods for the Examination of Water and Wastewater, 10th Edition. Washington, D.C. 1949.

Anderson K.L., Whitlock J.E., Harwood V.J. Persistence and differential survival of fecal indicator bacteria in subtropical waters and sediments. *Appl Environ Microbiol.* 2005, *71(6)*, 3041-8.

Anderson, I.C., Rhodes, M., Kator, H. Seasonal variation in survival of *Escherichia coli* exposed in situ in membrane diffusion chambers containing filters and nonfiltered estuarine water. *App. Env. Microbiol.* 1983, *456*, 1877-1883.

Anderson, I.C., Rhodes, M., Kator, H., Sublethal stress in *Escherichia coli*: a function of salinity. *Appl. Environ. Microbiol.* 1979, *38(6)*, 1147-1152.

Antich, A., Gobas, H.R., Salgaonkar, J. The Santa Monica Urban Runoff Recycling Facililty and the Sustainable Environment. *Stormwater – Journal of Surface Water Professionals*, 2003.

Association of Bay Area Governments (ABAG). Results from the Demonstration Urban Stormwater Treatment (DUST) Marsh. International Stormwater BMP Database. http://www.bmpdatabase.org. 2000.

Atherholt, T. *Microbial Source Tracking: Library Based Methods*. Trenton, New Jersey: Division of Science, Research, and Technology, 2004.

Atherholt, T.B., LeChevallier, M.W., Norton, W.D., Rosen, J.S. Effect of rainfall on *Giardia* and crypto. *J. Am. Water Works Assoc.* 1998, *909*, 66-80.

Auer, M.T., Niehaus, S.L Modeling Fecal Coliform Bacteria--I. Field and Laboratory Determination of Loss Kinetics. *Water Research* 1993, *27*(4), 693-701.

Bai S., Lung W.S. Modeling sediment impact on the transport of fecal bacteria. *Water Res.* 2005, *39(20)*, 5232-40.

Barcina, I., Lebaron, P., Vives-Rego, J. Survival of allochthonous bacteria in aquatic systems: A biological approach. *FEMS Microbiology Ecology* 1997, *23*, 1-9.

Barnes, B., Gordon, D.M. Coliform dynamics and the implications for source tracking. *Environ. Microbiol.* 2004, *6*, 501-509.

Barrett, M.E. *Complying with the Edwards Aquifer Rules: Technical Guidance on Best Management Practices,* Austin, Texas: Natural Resources Conservation Commission, 1999.

Barwick R.S., Levy, D.A., Craun, G.F., Beach, M.J., Calderon, R.L. Surveillance for waterborne disease outbreaks - United States, 1997-1998. *CDC Surveillance Summaries MMWR* 2000, *49(No. SS-4)*,1-35.

Baudart, J., Grabulos, J., Barusseau, J.P., Lebaron, P. *Salmonella spp.* and fecal coliform loads in coastal waters from a point vs. nonpoint source of pollution. *J. Environ. Qual.* 2000, *29*, 241-250.

Belas, R.M., Colwell, R.R. Adsorption kinetics of laterally and polarly flagellated vibrio. *J. Bacteriol.* 1982, *151*, 1568-1580.

Bernhard, A.E., Field, K.G. Identification of nonpoint sources of fecal pollution in coastal waters by using host-specific 16s ribosomal DNA genetic markers from fecal anaerobes. *Appl. Environ. Microbiol.* 2000, *66*, 1587-1594.

Bernstein, B.B., Griffith, J.F., Weisberg, S.B. *Microbiological Source Tracking Workshop: Summary of Proceedings,* Workshop on Microbial Source Tracking, Feb 5, 2002, Southern California Coastal Water Research Project (SCCWRP): Irvine, CA. 2002.

Berry, S.A., Noton, B.G. Survival of bacteriophages in seawater. *Water Res.* 1976, *10,* 323-327.

Bertrand-Krajewski, J.L., Chebbo, G., Saget, A. Distribution of pollutant mass vs. volume in stormwater discharges and the first flush phenomenon. *Water Res.* 1998, *32(8)*, 2341–2356.

Besnard, V., Federighi, M., Declercq, E., Jugiau, F., Cappelier, J.M. Environmental and physico-chemical factors induce VBNC state in *Listeria monocytogenes*. *Vet. Res.* 2002, *33*, 359-370.

Black, R.E., Levine, M.M., Clements, M.L., Hughes, T.P., Blaser, M.J. Experimental *Campylobacter* jejuni infection in humans. *J. Infect. Dis.*1988. 157:472-479.

Blackburn, B.G., Craun, G.F., Yoder, J.S., Hill, V., Calderon, R.L., Chen, N., Lee, S.H., Levy, D.A., Beach, M.J. Surveillance for waterborne-disease outbreaks associated with drinking water--United States, 2001-2002. *MMWR Surveill. Summ.* 2004, *538*, 23-45.

Boehm, A.B., Grant, S.B., Kim, J.H., Mowbray, S.L., Mcgee, C.D., Clark, C.D., Foley D.M., Wellman, D.E. Decadal and shorter period variability of surf zone water quality at Huntington Beach, California. *Environ. Sci. Technol.* 2002, *3618*, 3885-3892.

Bonjoch, X., Balleste, E., Blanch, A.R. Multiplex PCR with 16s RNA gene-targeted primers of *Bifidobacterium spp.* to identify sources of fecal pollution. *Appl. Eniviron. Microbiol.* 2004, *70,* 3171-3175.

Borden, R.C., Dorn, J.L., Stillman, J.B., Liehr, S.K. *Evaluation of Ponds and Wetlands for Protection of Public Water Supplies.* Raleigh, NC: Water Resource Research. Institute, North Carolina State University. 1996.

Borst, M., Selvakumar, A. Particle-associated microorganisms in stormwater runoff. *Wat. Res.* 2003, *37*, 215-223.

Boyer, D.G., Kuczynska, E. Storm and seasonal distributions of fecal coliforms and *Cryptosporidium* in a spring. *J. Amer. Water Res. Assoc.* 2003, *39 (6),* 1449-1456.

Brown, E., Caraco, D., Pitt, R. *Illicit Discharge Detection and Elimination: A Guidance Manual for Program Development and Technical Assessments*, Report for the U.S.Environmental Protection Agency, Ellicott City, MD: Office of Wastewater Management, Center for Watershed Protection, 2004.

Brown, W., Schueler, T. *The Economics of Stormwater BMPs in the Mid-Atlantic Region,* Report for Chesapeake Research Consortium, Ellicott City, MD: Center for Watershed Protection, 1997.

Bruce, I.D. Subsurface Flow Wetland Treatment of Dairy Farm Stormwater. *Proceedings from the 2003 Georgia Basin/Puget Sound Research Conference*. Vancouver, B.C., March 31-April 3, Droscher, T. W., Fraser, D.A., Eds., Olympia, WA: Puget Sound Action Team, 2003.

Brueske, C.C. *Technology Review: Ultra-Urban Stormwater Treatment Technologies.* Seattle, WA: Department of Civil and Environmental Engineering, University of Washington, 2000.

Buckley, R., Clough, E., Warnken, W., Wild, C. Coliform bacteria in streambed sediments in a subtropical rainforest conservation reserve. *Water Research* 1998, *32*(6), 1852-1856.

Burkhardt III, W., Calci, K.R., Watkins, D.W., Rippey, S. R., Chirtel, S. J. Inactivation of indicator microorganisms in estuarine waters. *Water Res.* 2000, *348(8)*, 2207-2214.

Burton, G.A., Pitt., R. *Stormwater Effects Handbook: A Toolbox for Watershed Managers, Scientists, and Engineers*. New York: CRC/Lewis Publishers, 2001.

Byappanahalli, M.N., Whitman, R. L, Shively, D.A., Sadowsky, M.J., Ishii, S. Population structure, persistence, and seasonality of autochthonous *Escherichia coli* in temperate, coastal forest soil from a Great Lakes watershed. *Environ. Microbiol.* 2006, *8(3)*, 503-513.

Cabelli, V.J., Levin, M.A., Dufour, A.P., McCabe, L.J. The Development of Criteria for Recreational Waters. In: International Symposium on Discharge of Sewage Fran Sea Outfalls. Ed. A.L.H. Gameson. Pergamon Press, London. 1975. p. 63.

Cabelli, V.J., Dufour, A.P., McCabe, L.J., Levin, M.A. Swimming-associated gastroenteritis and water quality. *Am. J. Epidemiol.* 1982, *115*, 606-616.

Cabelli, V.J. Public Health and Water Quality Significance of Viral Diseases Transmitted by Drinking Water and Recreational Water. In: *Water Virology*, W.O. Grabow, ed.; *Water Sci. Technol.* 1983a, *155*, 1-15.

Cabelli, V.J., Dufour, A.P., McCabe, L.J., Levin, M.A. A marine recreational water quality criterion consistent with indicator concepts and risk analysis. *J.Water Poll. Control Fed.* 1983b, *55(10),* 1306-14.

Calderon, R, Mood, E, Dufour, A. Health effects of swimmers and nonpoint sources of contaminated water. *Int J Environ Health Res.* 1991, *1*, 21-31.

California Department of Transportation (CalTrans). *BMP Retrofit Pilot Program*, Sacramento, CA: Division of Environmental Analysis, 2004.

California Stormwater Quality Association (CASQA). *Stormwater Best Management Practice BMP Handbook – Municipal.* Menlo Park, CA. 2003.

Carter, V.L., Andersen, M.E., Crawl, J.R., Culver, B.D., Hammond, J.W., Lewis, T.R., Lieberman, J., McDermott, F.T., Melvin, W.W., Murphy, M.K., Pagnotto, L.D., Ratney, TLVs: Threshold limit values for chemical substances in the work environment adopted by ACGIH for 1984-1985. 1983.

Center for Watershed Protection (CWP). Pollutant removal dynamics of three wet ponds in Canada. *Watershed Protection Techniques, Bull. Urban Wateshed Restor. Prot. Tools.* 2000, *3(3)*, 721-28.

Center for Watershed Protection (CWP). *Stormwater Practice Fact Sheets*, www.stormwatercenter.net, 2001

Chamberlain, C.E., Mitchell, R. A Decay Model for Enteric Bacteria in Natural Waters. In *Water Pollution Microbiology*, volume 2, R. Mitchell, Ed., New York: John Wiley and Sons, 1978, p. 325-348.

Chappell, C.L., Okhuysen, P.C., Sterling, C R. and DuPont, H.L. *Cryptosporidium parvum*: intensity of infection and oocyst excreta patterns in healthy volunteers. *J. Infect. Dis.* 1996, *173*, 232–236.

Chapron, C.D., Ballester, N.A., Fontaine, J.H., Frades, C.N., Margolin, A.B. Detection of astroviruses, enteroviruses, and adenovirus types 40 and 41 in surface waters collected and evaluated by the information collection rule and an integrated cell culture-nested PCR procedure. *Appl. Environ. Microbiol.* 2000, *66*, 2520–2525.

Characklis, G.W., Dilts, M.J., Simmons, O.D., III, Likirdopulos, C.A., Krometis, L.A., Sobsey, M.D. Microbial partitioning to settleable particles in stormwater. *Water Res.* 2005, *39(9)*, 1773-82.

Cheung, W. H., Chang, K.C., Hung, R.P., Kleevens, J. W. Health effects of beach water pollution in Hong Kong. *Epidemiol. Infect.* 1990, *105(1)*, 139-62.

Choi, S., Jiang, S.C. Real-time PCR quantification of human adenoviruses in urban rivers indicates genome prevalence but low infectivity. *Appl. Environ. Microbiol* 2005, *71*, 7426-7433.

Chung, H., Sobsey, M.D. Comparative survival of indicator viruses and enteric viruses in seawater and sediment. *Wat. Sci. Tech.* 1993, *27*, 425-428.

City of Austin. *Evaluation of Nonpoint Source Controls,* EPA/TNRCC Section 319, Grant Report. City of Austin, TX, 1997.

City of Laguna Niguel, County of Orange. *Clean-up & Abatement Order 99-211 18th Quarterly Progress Report.* City of Laguna Niguel, CA, 2004.

Clark C.S., Bjornson, A.B., Schiff,G.M., Phair, J.P., Van Meer, G.L., Gartside, P.S. Sewage worker's syndrome. *Lancet.* 1977, *1(8019)*, 1009.

Clark, C.S., Bjornson, H.S., Holland, J.W., Elia, V. J., Majeti, V. A., Meyer, C. R., Balistreri, W. F., Van Meer, G.L., Gartside, P.S., Specker, B.L., Linneman, C.C., Jaffa, R., et al. Evaluation of the health risks associated with the treatment and disposal of municipal wastewater and sludge, EPA 600/1-80-030, U.S. EPA, Health Effects Research Laboratory, PB81-175945, Washington D.C.: U.S. Government Printing Office, 1981.

Clark, C.S., Van Meer, G.L., Linneman, C. C., Bjornson, A. B., Gartside, P. S., Schiff, G. M., Trible, S.E., Alexander, D., Cleary, E.J., Phair, J. P. . *Health effects of occupational exposure to wastewater*, EPA 600/9-80-028, PB81-169864, Proc. Symp. Sept. 19-21, 1979, U.S. EPA, Office of Research and Development, Washington D.C.: U.S. Government Printing Office, 239-264, 1980.

Clark, C.S., Linneman, C.C., Gartside, P.S. Phair, J.P., Blacklow, N., Zeiss, C.R. Serologic survey of rotavirus, Norwalk agent and *Prototheca wickerhamii* in wastewater workers. *Am. J. Public. Health*. 1985, *75(1)*, 83-85.

Clausen, J.C., Belanger, P., Board, S., Dietz, M., Phillips, R., Sonstrom, R. *Stormwater Treatment Devices Section 319 Project: Project #99-07.* Hartford, CT: Connecticut Department of Environmental Protection. 2002.

Colford Jr., J.M., Tager, I., Byers, L.F., Ricci, P., Hubbard, A., Horner, R. Methods for assessing the public health impact of outflows from combined sewer systems. *J. Air Waste Manage. Assoc.* 1999, *494,* 454-62.

Colford, J.M., Jr., T.J. Wade, Colford, J.M,.Jr., Wade, T.J., Sandhu, S.K., Wright, C.C., Lee, S, Shaw, S., Fox, K., Burns, S., Benker, A., Brookhart, M.A., van der Laan, M., Levy, D.A. A randomized controlled trial of in-home drinking water intervention to reduce gastrointestinal illness. *Amer. J. Epidemiol*. 2005, *161(5),* 472-482

Cooper, R.C. *Disease Risk Among Sewage Plant Operators: A Review,* Report 91-1, Sanitary Engineering and Environmental Health Research Laboratory, 1991.

Cooper, R.C., Olivieri, A.W., Danielson, R.E., Badger, P.G. *Evaluation of military field-water quality. Vol. 6. Infectious organisms of military concern associated with nonconsumptive exposure: Assessment of health risks and recommendations for establishing related standards*, UCRL-21008, Vol. 5, 1986.

Corbett, S.J., Rubin, G.L., Curry, G.K., Kleinbaum, D.G. The health effects of swimming at Sydney beaches. *Amer. J. Public Health* 1993, *83(12)*, 1701-6.

Couch, R.B., Cate, T.R., Gerone, P.J., Fleet, W., Lang, D., Griffith, W., Knight, V. Production of illness with a small-particle aerosol of Coxsackie A21. *J. Clin. Invest.* 1965, *44*(4), 535–542.

Couch, R.B., Cate, T.R., Fleet, W.F., Gerone, P.J., and Knight, V. Aerosol-induced adenoviral illness resembling the naturally occurring illness in military recruits. *Am. Rev. Respir. Dis.* 1966. 93: 529–535.

Couch, R.B., Knight, V., Douglas, R.G., Black, S.H., Hamory, B.H. The minimal infectious dose of adenovirus Type 4; the case for natural transmission by viral aerosol. *Trans. Am. Clin. Climatol. Assoc.* 1969, *80*, 205–211.

Cox, C.R. Acceptable standards for natural waters used for bathing. Proceedings American Society of Civil Engineers 1951, 77, 74.

Crabill, C., Donald, R., Snelling, J., Foust, R., Southam, G. The impact of sediment fecal coliform reservoirs on seasonal water quality in Oak Creek, Arizona. *Wat. Res.* 1999, *339*, 2163-2171.

Craig, D.L., Fallowfield, H.J., Cromar, N.J.. Enumeration of faecal coliforms from recreational coastal sites: evaluation of techniques for the separation of bacteria from sediments. *J. Appl. Microbiol.*2002, *93*, 557-565.

Craig, D.L., Fallowfield, H.J., Cromar, N.J.. Use of microcosms to determine persistence of Escherichia coli in recreational coastal water and sediment and validation with in situ measurements. *J. Appl. Microbiol.* 2004, 96, 922-930.

Curriero, F.C., Patz, J.A., Rose, J.B., Lele, S. The association between extreme precipitation and waterborne disease outbreaks in the United States, 1948-1994. *Amer. J. Pub. Health.* 2001, *918,* 1194-1199.

Dai, X., Boll, J. Surface water quality: Evaluation of attachment of *Crytosporidium parvum* and *Giardia lamblia* to soil particles. *J. Environ. Qual.* 2003, *32*, 296-304.

Davies, C.M., Bavor, H.J. The fate of stormwater-associated bacteria in constructed wetland and water pollution control pond systems. *J. App. Microbiol.* 2000, *89*, 349-360.

Davies, C.M., Long, J.A.H., Donald, M., Ashbolt, N.J. Survival of fecal microorganisms in marine and freshwater sediments. *Appl. Environ. Microbiol.* 1995, *61(5),* 1888-1896.

Davies, C.M., Yousefi, Z., Bavor, H.J. Occurrence of coliphages in urban stormwater and their fate in stormwater management systems. *Letters Appl. Microbiol.* 2003, *37*, 299-303.

Davies-Colley, R.J., Donnison, A.M., Speed, D.J., Ross, C.M. Inactivation of faecal indicator micro-organisms in waste stabilisation ponds: Interactions of environmental factors with sunlight. *Water Res.* 1999, *335*, 1220-1230.

Desmarais, T.M., Solo-Gabriele, H.M., Palmer, C.J. Influence of soil on fecal indicator organisms in a tidally influenced subtropical environment. *Appl. Environ. Microbiol.* 2002, *683*, 1165-1172.

Dufour, A.P. *Health Effects Criteria for Fresh Recreational Waters*. Washington DC: Office of Research and Development, United States EPA. 1984.

Dupont, H.L., Chappell, C.L., Sterling, C.R., Okhuysen, P.C., Rose, J.B., Jakubowski, W. The infectivity of *Cryptosporidium parvum* in healthy volunteers. *New Engl. J. Med.* 1995, *332*(13), 855–859.

Dwight, R.H., Baker, D.B., Semenza, J.C., Olson, B.H. Health Effects Associated With Recreational Coastal Water Use: Urban Versus Rural California. *Amer. J. Publ. Health.* 2004, *94(4)*, 565-567.

Ebdon, J.E., Wallis, J.L., Taylor, H.D. A simplified low-cost approach to antibiotic resistance profiling for faecal source tracking. *Water Sci. Technol.* 2004, *150*, 185-191.

Eisenberg, J.N.S., Seto, E.Y.W., Colford, J.M., Jr., Olivieri, A., Spear, R.C. An analysis of the Milwaukee cryptosporidiosis outbreak based on a dynamic model of the infection process. *Epidemiol.* 1998, *9(3)*, 255-263.

Emmerling-Dinovo, C. Stormwater detention basins and residential locational decisions. *Water Res. Bull.* 1995, *31(3)*, 515-521.

Englehardt, J. Predictive Bayesian dose-response assessment for appraising absolute health risk from available information. *Human Ecol. Risk Assess*. 2004, *10*(1), 69–74.

Enriquez, C.E., Hurst, C.J., Gerba, C.P. Survival of the enteric adenoviruses 40 and 41 in tap, sea, and waste water. *Water Res.* 1995, *2911*, 2548-2553.

Enzinger, R.M., Cooper, R.C. Role of bacteria and protozoa in the removal of *Escherichia coli* from estuarine waters. *Appl. Environ. Microbiol.* 1976, *315*, 758-763.

Evison, L.M. Comparative studies on the survival of indicator organisms and pathogens in fresh and sea water. *Wat. Sci. Tech.* 1988, *20*, 309-315.

Fankhauser, R.L., Noel, J.S., Monroe, S.S., Ando, T., Glass, R.I. Molecular epidemiology Of Norwalk-like viruses in outbreaks of gastroenteritis in the United States. *J. Infect. Dis.* 1998, *1786*, 1571-1578.

Fattal, B., Peleg-Olevsky, E., Yoshpe-Purer, Y., Shuval, H. The association between morbidity among bathers and microbial quality of seawater. *Wat Sci Tech.* 1986, *18*, 59-69.

Fayer, R., Graczyk, T.K., Lewis, E.J., Trout, J.M., Farley, C.A. Survival of infectious *Cryptosporidium parvum* oocysts in seawater and eastern oysters (*Crassostrea virginics*) in the Chesapeake Bay. *Appl. Environ. Microbiol.* 1998, *643(3)*, 1070-1074.

Ferguson, C.M., Coote, B.G., Ashbolt, N.J., Stevenson, I.M. Relationships between Indicators, pathogens and water quality in an estuarine system. *Water Res.* 1996, *30(9)*, 2045-2054.

Ferguson, D.M., Moore, D.F., Getrich, M.A., Zhowandai, M.H. Enumeration and speciation of enterococci found in marine and intertidal sediments and coastal water in southern California. *Applied Microbiology* 2005, *99*, 598-608.

Ferley, J.P., Zmirou, D., Balducci, F., Baleux, B., Fera, P., Larbaigt, G., Jacq, E., Moissonnier, B., Blineau, A., Boudot, J. Epidemiological significance of microbiological pollution criteria for river recreational waters. *Intern. J. Epidemiol.* 1989, *181*, 198-205.

Fewtrell., L., Godfree, A.F., Jones, F., Kay, D., Salmon, R.L., Wyer, M.D. Health effects of white water canoeing. *Lancet.* 1992, *339*, 1587-1589.

Fleisher, J.M., Jones, F., Kay, D., Stanwell-Smith, R., Wyer, M., Morano, R. Water and non-water related risk factors for gastroenteritis among bathers exposed to sewage-contaminated marine waters. *Intern. J. Epidemiol.* 1993, *224*, 698-708.

Fleisher, J.M., Kay, D., Wyer, M.D., Godfree, A.F. Estimates of the severity of illnesses associated with bathing in marine recreational waters contaminated with domestic sewage. *Intern. J. .Epidemiol.* 1998, *27(4), 722-726*

Foulon, G., Maurin, J., Quoi, N., Martin –Buoyer, G. Etude de la morbidite humaine en relation avec la pollutinon bacteriologique des eaux de baignade en mer. *Rev Fr Sci Eau* 1983, 2, 127-143.

Frost, W.H. Report of committee on sanitary control of the shellfish industry in the United States. Public Health Reports Supplement. 1925. 53, 1-17.

Fujioka, R.S. Monitoring coastal marine waters for spore-forming bacteria of faecal and soil origin to determine point from non-point source pollution. *Wat. Sci. Tech.* 2001, *447*, 181-188.

Fujioka, R.S. Yoneyama, B.S. Sunlight inactivation of human enteric viruses and fecal bacteria. *Water Sci. Technol.* 2002, *46,* 291-295.

Fujioka, R.S., Hashimoto, H.H., Siwak, E.B., Young, R.H. Effect of sunlight on survival of indicator bacterial in seawater. *Appl. Environ. Microbiol.* 1981, *413*, 690-696.

Fujioka, R.S., Loh, P.C., Lau, L.S. Survival of human enteroviruses in the Hawaiian ocean environment: Evidence for virus-inactivating microorganisms. *Appl. Environ. Microbiol.* 1980, *396*, 1105-1110.

Gamo, M. Shoji, T. A method of profiling microbial communities based on a most-probable-number assay that uses biolog plates and mutliple sole carbon sources. *Appl. Eniviron. Microbiol.* 1999, *65*, 4419-4424.

Gantzer, C., Quignon, F., Schwartzbrod, L. Poliovirus-1 adsorption onto and desorption from montmorillonite in seawater – survival of the absorbed virus. *Environ. Tech.* 1994, *153*, 271-278.

Garber, W.F. Bacteriological standards for bathing waters. *Sewage & Industrial Wastes* 1956, 28(6), 795-807.

Gauthier, M.J., Munro, P.M., Breittmayer, V. A. Influence of prior growth conditions on low nutrient response on *Escherichia coli* in seawater. *Can. J. Microbiol.* 1989, *35,* 379-383.

Gauthier, M.J., Flatau, G.N., Le Rudulier, D., Clément, R.L., Combarro, M.P. Intracellular accumulation of potassium and glutamate specifically enhances survival of *Escherichia coli* in seawater. *Appl. Environ. Microbiol.* 1991, *57(1)*, 272-276.

Gerba, C.P., McLeod, J.S. Effect of sediments on the survival of *Escherichia coli* in marine waters. *Appl. Environ. Microbiol.* 1976, *321*, 114-120.

Gibson, C.J., Stadterman, K.L., States, S., Sykora, J. Combined sewer overflows: A source of *Cryptosporidium* and *Giardia*? *Wat. Sci. Tech.* 1998, *3812*, 67-72.

Goldberg, D. *Dayton Avenue Swale Biofiltration Study.* Seattle, WA: Seattle Engineering Dept., 1993.

Gonzalez, J.M. Suttle, C.A. Grazing by marine nanoflagellates on viruses and virus-sized particles: Ingestion and digestion. *Mar. Ecol. Prog. Ser.* 1993, *94*,1-10.

Goyal, S.M., Gerba, C.P., Melnick, J.L. Occurrence and distribution of bacterial indicators and pathogens in canal communities along the Texas coast. *Appl. Environ. Microbiol.* 1977, *34(2)*, 139-149.

Griffith, J.F., Weisberg, S.B., Mcgee, C.D. Evaluation of microbial source tracking methods using mixed fecal sources in aqueous test samples. *J. Wat. Health.* 2003, *01(4)*, 141-151.

Grimes, D.J. Bacteriological water quality effects of hydraulically dredging contaminated upper Mississippi River bottom sediment. *Appl Environ Microbiol.* 1980, *39(4)*, 782-789.

Haas, C.N. Estimation of risk due to low doses of microorganisms: A comparison of alternative methodologies. *Amer. J. Epidemiol.* 1983, *55*, 573-582.

Haas, C.N., Rose, J.B., Gerba, C.P., Quantitative microbial risk assessment. J.W. Wiley, Inc. 1999.

Hagedorn, C., Robinson, S.L., Filtz, J.R., Grubs, S.M., Angier, T.A., Reneau, R.B. Determining sources of fecal pollution in a rural virginia watershed with antibiotic resistance patterns in fecal streptococci. *Appl. Eniviron. Microbiol.* 1999, *65*, 5522-5531.

Haile, R.W., Witte, J.J., Gold, M., Cressey, R., Mcgee, C., Millikan, R.C., Glasser, A., Harawa, N., Ervin, C., Harmon, P., Harper, J., Dermand, J., Alamillo, J., Barrett, K., Nides, M., Wang, G.-Y. The health effects of swimming in ocean water contaminated by storm drain runoff. *Epidemiol.* 1999, *104*, 355-363.

Halton, J.E., Nehlsen, W.R. Survival of *Escherichia coli* in zero-degree centigrade sea water. *J. Water Pollut. Control Fed.* 1968, *40(5),* 865-8.

Hanes, N.B., Fragala, R., Effect of seawater concentration on survival of indicator bacteria. *J. Water Pollut. Control Fed.* 1967, *39*, 97-104.

Hardina, C.M., Fujioka, R.S. Soil: the environmental source of *Escherichia coli* and *Enterococci* in Hawaii's streams. *Environ, Toxicol. Water Qual.* 1991, *6*, 185-195.

Harper, H.H., Herr, J.L., Baker, D. *Performance Evaluation of Dry Detention Stormwater Management Systems.* Orlando, FL: Environmental Research & Design Inc., 1999.

Hartel, P.G., Summer, J.D., Hill, J.L., Collins, J.V., Entry, J.A., Segars, W.I. Geographic variability of *Escherichia coli* ribotypes from animals in Idaho and Georgia. *J. Environ. Qual.* 2002, *31*, 1273-1278.

Harwood, V.J., Wiggins, B., Hagedorn, C., Ellender, R.D., Gooch, J., Kern, J., Samadpour, M., Chapman, A.C.H., Robinson, B.J., Thompson, B.C. Phenotypic library-based microbial source tracking methods: Efficacy in the California collaborative study. *J. Wat. Health.* 2003, *01(4),* 153-166.

Henderson, J.M. Enteric Disease Criteria for Recreational Waters. *J. San. Eng. Div.* 1968, 94, 1253.

Hendricks, C.W. Enteric bacterial metabolism of stream sediment eluates. *Can. J. Comp. Med.* 1971, *17*, 551-556.

Hilding, K. Longevity of infiltration basins assessed in Puget Sound. *Water. Protect. Tech.* 1996, 1(3):124–125.

Hinds, T.D., Rachel, J.R., Brown, R., Burns, E.H.J. *Reduction of Fecal Coliform Levels in Two Created Wetlands at the Olentangy River Wetland Research Park,* Portsmouth, OH: Dept. Natural Sciences, Shawnee State University, 2004.

Hoppin, J. Ryan, P.B., Graham, J.D. *Risk Assessment in the Federal Government: Questions and Answers,* Boston, MA: Center for Risk Analysis, Harvard School of Public Health, 1993.

Hornick, R.B., Music, S.I., Wenzel, R., Cash, R., Libonati, J.P., Woodward, T.E. The Broad Street pump revisited: response of volunteers to ingested cholera vibrios. *Bull. NY Acad. Med.* 1971, *47*(10), 1181–1191.

Howell, J.M., Coyne, M.S, Cornelius, P.L. Effect of sediment particle size and temperature on fecal bacteria mortality rates and the fecal coliform/fecal streptococci ratio. *J. Environ. Qual.* 1996, *25*, 1216-1220.

Hughes, K.A. Influence of seasonal environmental variables on the distribution of presumptive fecal coliforms around an Antarctic research station. *Appl. Environ. Microbiol.* 2003, *698*, 4884-4891.

Hulton, C. S.J., Higgins, C.F., Sharp, P.M. Eric sequences: A novel family of repetitive elements in the genome of *Escherichia coli, Salmonella typhimurium* and other enteric bacteria. *Mol. Microbiol.* 1991, *5*, 825-834.

Hunter, P.R. Drinking water and diarrhoeal disease due to *E. coli*. *J. Water Health*. 2003, *1(2)*, 65-72.

Iftimovici, R, Iacobescu, V, Copelovici, Y, Dinca, A, Iordan, L, Niculescu, R, Teleguta L, Chelaru, M. Prevalence of antiviral antibodies in workers handling wastewater and sludge. *Virol.* 1980, *31(3)*, 187-89.

International Life Sciences Institute (ILSI). A conceptual framework to assess the risks of human disease following exposure to pathogens. *Risk Anal.* 1996, *16(6),* 841:848.

International Life Sciences Institute (ILSI). Revised Framework for Microbial Risk Assessment. ILSI Risk Science Workshop Report. 2000. ILSI Press, Washington, D.C.

Izbicki, J.A., Pimentel, M.I., Leddy, M., Bergamaschi, B. Microbial and Dissolved Organic Carbon Characterization of Stormflow in the Santa Ana River at Imperial Highway, Southern California, 1999-2002; Sacramento, California, 2004.

Jamieson, R.C., Gordon, R.J., Sharples, K.E., Stratton, G.W., Madani, A. Movement and persistence of fecal bacteria in agricultural soils and subsurface drainage water: a review. *Can. Biosyst. Engin*. 2002, *44,* 1.1-1.9.

Jamieson, R.C., Joy, D.M., Lee, H., Kostaschuk, R., Gordon, R.J. Persistence of enteric bacteria in alluvial streams. *J. Environ. Eng. Sci*. 2004, *3(3)*, 203-212.

Jamieson R., Joy, D.M., Lee, H., Kostaschuk, R., Gordon, R. Transport and deposition of sediment-associated Escherichia coli in natural streams. *Water Res*. 2005, *39*(12), 2665-75.

Jiang, S., Chu, W. PCR detention of pathogenic viruses in Southern California urban rivers. *J. Appl. Microbiol.* 2004, *97*, 17-28.

Jiang, S., Noble, R., Chu, W. Human adenoviruses and coliphages in urban runoff-impacted coastal waters of Southern California. *Appl. Environ. Microbiol.* 2001, *671*, 179-184.

Jin, G., Englande, A.J., Bradford, H., Jeng, H.-W. Comparison of *E. coli*, enterococci, and fecal coliform as indicators for brackish water quality assessment. *Wat. Environ. Res.* 2004, *76*, 245-255.

Johnson, D.C., Enriquez, C.E., Pepper, I.L., Davis, T.L., Gerba, C.P., Rose, J.B. Survival of *Giardia*, *Cryptosporidium*, poliovirus, and *Salmonella* in marine waters. *Wat. Sci. Tech.* 1997, *35*, 261-268.

Johnson, J.M., Weagant, S.D., Jinneman, K.C., Bryant, J.L. Use of pulsed field gel electrophoresis for epidemiological study of *Escherichia coli* 0157:H7 during a food-borne outbreak. *Appl. Environ. Microbiol.* 1995, *61,* 2806-2808.

Johnson, L.K., Brown, M.B., Carruthers, E.A., Ferguson, J.A., Dombek, P.E., Sadowsky, M.J. Sample size, library composition, and genotypic diversity among natural populations of *Escherichia coli* from different animals influence accuracy of determining sources of fecal pollution. *Appl. Environ. Microbiol.* 2004, *70*, 4478-4485.

Jones, S.H., Langan, R. *Assessment of the Effectiveness of Permanent Stormwater Control Measures,* Durham, NH: University of New Hampshire. 1996.

Kannenberg, E., Carlson, L., Russell, W. Lipid A and O-chain modifications cause Rhizobium lipopolysaccharides to become hydrophobic during bacteroid development. *Molecular Microbiology* 2001, *39*(2), 379-392.

Katz, M., Plotkin, S.A. Minimal infective dose of attenuated poliovirus for man. *Am. J. Public Health Nations Health.* 1967, *57*(10), 1837–1840.

Kay, D., Fleisher, J.M., Salmon, R.L., Jones, F., Wyer, M.D., Godfree, A.F., Zelenauch-Jacquotte, Z., Shore, R. Predicting likelihood of gastroenteritis from sea bathing: results from randomized exposure. *Lancet.* 1994, *344 (8927)*, 905-909.

Khuder, S.A., Arthur, T., Bisesi, M.S., Schaub, E.A. Prevalence of infectious diseases and associated symptoms in wastewater treatment workers. *Am. J. Ind. Med.* 1998, *33(6)*, 571-577.

Kistemann, T., Claben, T., Koch, C., Dangendorf, F., Fischeder, R., Gebel, J., Vacata, V., Exner, M. Microbial load of drinking water reservoir tributaries during extreme rainfall and runoff. *Appl. Environ. Microbiol.* 2002, *685*, 2188-2197.

Knobloch, J., Bialek, R., Hasemann, J. Intestinal protozoal infestation in persons with occupational sewage contact. *Dtsch. Med. Wochenschr.* 1983, *108*, 57-60.

Koopmans, M., Vinje, J., Duizer, E., de Wit, M., van Duijnhoven, Y. Molecular epidemiology of human enteric caliciviruses in The Netherlands. *Gastroent. Vir.* 2001, *238*, 197-218.

Kueh, C.S.W., Grohmann, G.S. Recovery of viruses and bacteria in waters off Bondi Beach: a pilot study. *Med. J. Austral.* 1989, *151*, 632-638.

Kueh, C.S.W., Tam, T.Y., Lee,T., Wong, S.L., Lloyd, O.L., Yu, I. T.S., Wong, T.W., Tam, J.S., Bassett, D.C.J. Epidemiological study of swimming-associated illnesses relating to bathing-beach water quality. *Water Sci. Tech.* 1995, *315-6*, 1-4.

Kurz, R.C. *Removal of Microbial Indicators from Stormwater Using Sand Filtration, Wet Detention, and Alum Treatment Best Management Practices.* Tampa, FL: Southwest Florida Water Management District. 1999.

Labelle, R.L., Gerba, C. Influence of pH, salinity, and organic matter on the adsorption of enteric viruses to estuarine sediment. *Appl. Environ. Microbiol.* 1979, *381*, 93-101.

Labelle, R.L., Gerba, C.P. Influence of estuarine sediment on virus survival under field conditions. *Appl. Environ. Microbiol.* 1980, *394*, 749-755.

Labelle, R.L., Gerba, C.P., Goyal, S.M., Melnick, J.L., Cech, I., Bogdan, G.F. Relationships between environmental factors, bacterial indicators, and the occurrence of enteric viruses in estuarine sediments. *Appl. Environ. Microbiol.* 1980, *393*, 588-596.

LaLiberte, P., Grimes, D.J. Survival of *Escherichia coli* in lake bottom sediment. *Appl. Environ. Microbiol.* 1982, *43(3)*, 623-628.

Le Guyader, F., Dubois, E., Menard, D., Pommepuy, M. Detection of hepatitis A virus, rotavirus, and enterovirus in naturally contaminated shellfish and sediment by reverse transcription-seminested PCR. *Appl. Environ. Microbiol.* 1994, *60(10)*, 3665-71.

Lee, J.V., Dawson, S.R., Ward, S., Surman, S.B., Neal, K.R. Bacteriophages are a better indicator of illness rates than bacteria amongst users of a white water course fed by a lowland river. *Water Science and Technology* 1997, *35* (11-12), 165-170.

Lee, S.H., Levy, D.A., Craub, G.F., Beach, M.J., Calderon, R.L. Surveillance for waterborne disease outbreaks: United States, 1999-2000. *Morb. Mortal. Wkly. Rep.* 2002, *51*, 1-45.

Lee, H.K., Jeong, Y.S. Comparison of total culturable virus assay and multiplex integrated cell culture-PCR for reliability of waterborne virus detection. *Applied and Environmental Microbiology* 2004, *70*, 3632-3636.

Lemarchand, K., Lebaron, P. Occurrence of *Salmonella spp.* and *Cryptosporidium spp.* in a French coastal watershed: relationship with fecal indicators. *FEMS Microbiology Letters.* 2003, *218*, 203-209.

Lepow, M.L., Warren, R.J., Ingram, V.G., Daugherty, S.C., Robbins, F.C. Sabin type I (LSc2ab) oral poliomyelitis vaccine. Effect of dose upon response of newborn infants. *Am. J. Dis. Child.* 1962, *104*, 67–71.

Levine, W.C., Stephenson, W.T. Waterborne disease outbreaks, 1986-1988. *Morb. Mort. Wkly. Rpt.* 1990, *39(1)*, 1-13.

Lightfoot, N. A prospective study of swimming related illness at six freshwater beaches in southern Ontario. Ph.D. Dissertation,. University of Toronto, Toronto, Canada 1989.

Linnemann, C.C., Jaffa, R., Gartside, P.S. Risk of infection associated with a wastewater spray irrigation system used for farming. *J. Occup. Med.* 1984, *1(26)*, 41-44

Lipp, E.K., Farrah, S.R., Rose, J.B. Assessment and impact of microbial fecal pollution and human enteric pathogens in a coastal community. *Mar. Poll. Bull.* 2001a, *42*, 286-293.

Lipp, E.K., Kurz, R., Vincent, R., Rodriguez-Palacios, C., Farrah, S.R., Rose, J.B. The Effects of Seasonal Variability and Weather on Microbial Fecal Pollution and Enteric Pathogens in a Subtropical Estuary. *Estuaries.* 2001b, *24(2)*, 266–276.

Lo, S., Gilbert, J.B., Hetrick, F. Stability of human enteroviruses in estuarine and marine waters. *Appl. Environ. Microbiol.* 1976, *322*, 24-249.

Loge, F.J., Thompson, D.E., Call, D.R. PCR detection of specific pathogens in water: A risk-based analysis. *Environ. Sci. Tech.* 2002, 36, 2754-2759.

Long, S.C., Mahar, E.J., Pei, R., Arango, C., Shafer, E., Schoenber, T.H. *Development of Source-Specific Indicator Organisms for Drinking Water,* Denver, CO: AWWA Research Foundation, 2002.

Macek, M., Carlos, G., Memije, P., Ramirez, P. Ciliate-vibrio cholerae interactions within a microbial loop: an experimental study. *Aquat. Microb. Ecol.* 1997, *13*, 257-266.

Maestre, A., Pitt, R. *The National Stormwater Quality Databases, Version 1.1 - A Compilation and Analysis of NPDES Stormwater Monitoring Information.* Tuscaloosa, AL: Dept. of Civil and Environmental Engineering, University of Alabama, 2005.

Mahler, B.J., Personne, J.-C., Lods, G.F., Drogue, C. Transport of free and particulate-associated bacteria in karst. *J. Hydrol.* 2000, *238*, 179-193.

Maier, R.M., Pepper, I.L., Gerba, C.P. *Environmental Microbiology*, San Diego, CA: Academic Press, 2000.

Maki, R.P., Hicks, R.E. Surface water quality: *Salmonella typhimurium* survival and viability is unaltered by suspended particles in freshwater. *J. Environ.Qual.* 2002, *31*, 1702-1709.

Mallin, M.A., Williams, K.E., Esham, E.C., Lowe, P. Effect of human development on bacteriological water quality in coastal watersheds. *Ecol. Appl.* 2000, *104*, 1047-1056.

Mallin, M.A., Ensign, S.H., Wheeler, T.L., Mayes, D.B. Pollutant removal efficacy of three wet detention ponds. *J. Environ. Qual.* 2002, *31*, 654–660.

Marino, F., Morinigo, M., Martinez-Manzanares, E., Borrego, J. Microbiological-Epidemiological study of selected marine beaches in Malaga (Spain). *Wat Sci Tech* 1995, *31*(5-6), 5-9.

Marino, R.P., Gannon, J.J. Survival of fecal coliforms and fecal streptococci in storm drain sediment. *Wat. Res.* 1991, *259*, 1089-1098.

Marshall, K.C. Clay mineralogy in relation to survival of soil bacteria. *Ann.Rev.* 1975, *13*, 357-373.

Mattsby, I, Rylander, R. Clinical and immunological findings in workers exposed to sewage dust. *J. Occup. Med.* 1978, *20*, 690–692.

McBride, G., Till, D., Ryan, T. *Pathogen Occurrence and Human Health Risk Assessment Analysis*, Freshwater Microbiology Research Programme Report, Ministry of the Environment, Health, Agriculture and Forestry, New Zealand, 2002.

McBride, G.B., Salmondo, C.E., Bandaranayake, D.R., Turner, S.J., Lewis, G.D., Till, DG. Health effects of marine bathing in New Zealand. *Int. J. Environ. Health Res.* 1998, *8*, 173-189.

McCambridge, J., McMeekin, T.A. Effect of solar radiation and predacious microorganisms on survival of fecal and other bacteria. *Appl. Environ. Microbiol.* 1981, *415*, 1083-1087.

McCambridge, J., McMeekin, T.A. Relative effects of bacterial and protozoan predators on survival of *Escherichia coli* in estuarine water samples. *Appl. Environ. Microbiol.* 1980, *405*, 907-911.

McCullough, N.B. and Eisele, C.W. Experimental human salmonellosis. I. Pathogenicity of strains of *Salmonella* meleagridis and *Salmonella* anatum obtained from spray dried whole egg. *J. Infect. Dis.* 1951, *88*, 278–289.

McLellan, S.L., Daniels, A.D., Salmore, A.K. Genetic characterization of *Escherichia coli* populations from host sources fecal pollution by using DNA fingerprinting. *Appl. Eniviron. Microbiol.* 2003, *69*, 2587-2594.

Mead, P.S., Slutsker, L, Dietz, V., Mccaig, L.F., Bresee, J.S., Shapiro, C., Griffin, P.M., Tauxe, R.V. Food related illness and death in the United States. *Emerg. Infect. Dis.* 1999, *55*, 607-625.

Medema, G., Van Asperen, I., Klokman-Houweling, J., Nooitgedagt, A., Van de Laar, M. The relationship between health effects in triathletes and microbiological quality of freshwater. *Water Sci. Technol.* 1995, *31*, 19-26.

Menon, P., Billen, G., Servais, P. Mortality rates of autochthonous and fecal bacteria in natural aquatic ecosystems. *Wat. Res.* 2003**,** *37*, 4151-4158.

Messner, M.J., Chappell, C.L., Okhuysen, P.O. Risk assessment for *Cryptosporidium*: a hierarchical Bayesian analysis of human dose response data. *Wat. Res.* 2001, *35*(16), 3934–3940.

Minor, T.E., Allen, C.I., Tsiatis, A.A., Nelson, D.B., D'Alessio, D.J. Human infective dose determinations for oral poliovirus type 1 vaccine in infants. *J. Clin. Microbiol.* 1981, *13*(2), 388–389.

Moore, B. The Case Against Microbial Standards for Bathing Beaches. In: International Symposium on Discharge of Sewage from Sea Outfalls. Ed. A.L.H. Gameson. Pergamon Press, London. 1975. p. 103.

Morris, R.D., Naumova, E.N., Griffiths, J.K. Did Milwaukee experience waterborne cryptosporidiosis before the large documented outbreak in 1993? *Epidemiol.* 1998, *9(3)*, 264-70.

Morris, R.D., Naumova, E.N., Levin, R., Munasinghe, R.L. Temporal variation in drinking water turbidity and diagnosed gastroenteritis in Milwaukee. *Am. J. Public Health.* 1996, *86(2)*, 237-9.

Moss, D.M., Chappell, C.L., Okhuysen, P.C., DuPont, H.L., Arrowood, M.J., Hightower, A.W., Lammie, P.J. The antibody response to 27-, 17-, and 15-kDa *Cryptosporidium* antigens following experimental infection in humans. *J. Infect. Dis.* 1998, *178*, 827–833.

Mubiru, D.N., Coyne, M.S, Grove, J.H. Mortality of *Escherichia coli* O157:H7 in two soils with different physical and chemical properties. *J. Environ. Qual*. 2000, *29*, 1821-1825.

Myoda, S.P., Carson, C.A., Fuhrmann, J. J., Hahm, B. K., Hartel, P. G., Yampara-Iquise, H., Johnson, L., Kuntz, R.L., Nakatsu, C. H., Sadowsky, M. J., Samadpour, M. Comparison of geotypic-based microbial source tracking methods requiring a host origin database. *J. Wat. Health*. 2003, *01(4),* 167-180.

Nasser, A.M., Zaruk, N., Tenenbaum, L., Netzan, Y. Comparative survival of *Cryptosporidium*, coxsackievirus A9 and *Escherichia coli* in stream, brackish and sea waters. *Wat. Sci. Tech*. 2003, *473*, 91-96.

National Research Council (NRC). Drinking Water and Health, Volume 1. Washington, D.C. National Academy Press. 1972.

National Research Council (NRC). *Risk Assessment in the Federal Government: Managing the Process,* Washington, D.C.: National Academy Press, 1983

National Research Council (NRC). Indicators for Waterborne Pathogens. Washington, D.C., National Academy Press. 2004.

Naumova, E.N., Egorov, A.I., Morris, R.D., Griffiths, J.K. The elderly and waterborne *Cryptosporidium* infection: Gastroenteritis hospitalizations before and during the 1993 Milwaukee outbreak. *Emerg. Infect. Dis.* 2003, *94*, 418-25.

Niemi, M. Survival of *Escherichia coli* phage T7 in different water types. *Wat. Res.* 1975, *10*, 751-755.

Noble, R.T., Allen, S.M., Blackwood, A.D., Chu, W., Jing, S.C., Lovelace, G.L., Sobsey, M.D., Stewart, J.R., Wait, D.A. Use of viral pathogens and indicators to differentiate between human and non-human fecal contamination in a microbial source tracking comparison study. *J. Wat. Health.* 2003, *01(4)*, 195-207.

Noble, R.T., Lee, I.M., Schiff, K.C. Inactivation of indicator micro-organisms from various sources of faecal contamination in seawater and freshwater. *J. Appl. Microbiol.* 2004, *96*, 464-472.

Novitsky, J.A. Heterotrophic activity throughout a vertical profile of seawater and sediment in Halifax Harbor, Canada. *Appl. Environ. Microbiol.* 1983, *456*, 1753-1760.

Nystrom, T. Stationary-phase physiology. *Ann. Rev. Microbiol.* 2004, *58*, 161-181.

Obiri-Danso, K., Jones, K. Distribution and seasonality of microbial indicators and thermophilic *Campylobacter*s in two freshwater bathing sites on the River Lune in Northwest England. *J. Appl. Microbiol.* 1999, *87*, 822-832.

Obiri-Danso, K., Jones, K. Intertidal sediments as reservoirs for hippurate negative *Campylobacter*s, *Salmonella*e and faecal indicators in three EU recognised bathing waters in North West England. *Wat. Res.* 2000, *342*, 519-527.

Obiri-Danso, K., Jones, K. The effects of UV and temperature on the survival of natural populations and pure cultures of *Campylobacter jejuni, C. coli, C. lari* and urease-positive thermophilic *Campylobacter*s (UPTC) in surface waters, *J. Appl. Microbiol.* 2001, *90*, 256-267.

O'Brien, R.T., Newman, J.S. Inactivation of polioviruses and coxsackieviruses in surface water. *Appl. Environ. Microbiol.* 1977, *33(2)*, 334-340

Oguma, K., Katayama, H., Mitani, H., Morita, S., Hirata, T., Ohgaki, S. Determination of pyrimidine dimers in *Escherichia coli* and *Cryptosporidium parvum* during UV light inactivation, photoreactivation, and dark repair. *Appl. Environ. Microbiol.* 2001, *67(10)*, 4630-4637.

Okhuysen, P.C., Chappell, C.L., Sterling, C.R., Jakubowski, W., DuPont, H.L. Susceptibility and serologic response of healthy adults to reinfection with *Cryptosporidium parvum*. *Infect. Immun.* 1998, *66*(2), 441–443.

Okhuysen, P.C., Chappell, C.L., Crabb, J.H., Sterling, C.R., DuPont, H.L. Virulence of three distinct *Cryptosporidium parvum* isolates for healthy adults. *J. Infect. Dis.* 1999 *180*, 1275–1281.

Olivieri, V.P., Kruse, C.W., Kawata, K., Smith, J.E. *Microorganisms in Urban Stormwater*, U.S. EPA 600/2-77-087, Washington, D.C.: U.S. Government Printing Office, 1977.

O'Shea, M.L., Field, R. Detection and disinfection of pathogens in storm-generated flows. *Can. J. Microbiol.* 1992, *38*, 267-276.

Oshiro, R., Fujioka, R. Sand, soil and pigeon droppings: sources of indicator bacteria in the waters of Hanauma Bay, Oahu, Hawaii. *Water Science and Technology* 1995, *31*, 251-254.

Paerl, H.W. Microbial attachment to particles in marine and freshwater ecosystems. *Microb. Ecol.* 1975, *2*, 73-83.

Palmateer, G.A., Dutka, B.J., Janzen, E.M., Meissner, S.M., Sakellaris, M.G. Coliphage and bacteriophage as indicators of recreational water quality. *Water Res.* 1991, *253*, 355-357.

Palmateer, G., McLean, D.E., Kutas, W.L., Meissner, S.M., Suspended particulate/bacterial interaction in agricultural drains. In: *Particulate Matter and Aquatic Contaminants*, S.S. Rao(Ed), Chelsea, MI: Lewis Publishers, 1993. pp. 1-40.

Parveen, S., Portier, K.M., Robinson, K., Edminston, L., Tamplin, M. Discriminate analysis of ribotypes of *Escherichia coli* for differentiating human and nonhuman sources of fecal pollution. *Appl. Environ. Microbiol.* 1999, *65*, 3132-3147.

Payment, P., Richardson, L., Siemiatycki, J., Dewar, R., Edwardes, M., Franco, E. A randomized trial to evaluate the risk of gastrointestinal disease due to consumption of drinking water meeting current microbiological standards. *Am. J. Public Health*. 1991, *81(6)*, 703-708.

Payment, P., Siemiatycki, J., Richardson, L., Gilles, R., Franco, E., Prevost, M. A prospective epidemiological study of gastrointestinal health effects due to the consumption of drinking water. *Int. J. Environ. Health Res*. 1997. *7*, 5-31.

Perdek, J.M., Arnone, R.D., Stinson, M. K. *Managing Urban Watershed Pathogen Contamination*, EPA/600/R-03/111, Tuccillo, M.E. (ed), US Environmental Protection Agency, Edison, NJ, Washington D.C. U.S.: Government Printing Office, 2003.

Philipp, R., Evans, E.J., Hughes, A.O., Grisdale, S.K., Enticott, R.G., Jephcott, A.E. Health risks of snorkel swimming in untreated water. *Int. J. Epidemiol*. 1985, *14*, 624-627.

Pike, E. *Health Effects of Sea Bathing*, WMI 9021, Phase III--Final Report to the Department of the Environment, WRC Report Number: DoE 3412/2. Medmenham, UK: Water Research Centre, 1994

Pitt, R.E. *Stormwater Detention Pond Design for Water Quality Management* (Draft). Lewis Publishers. 1994.

Pitt, R.E.; Field, R. Urban Storm-induced Discharge Impacts: US Environmental Protection Agency Research Program Review. *Water Sci. Tech.* 1990, *22(10/11)*, 1-7.

Pitt, R., Lalor, M., Field, R., Adrian, D.D., Barbe, D. *Investigation of Inappropriate Pollutant Entries into Storm Drainage Systems, a Users Guide,* EPA/600/r-921238, United States Environmental Protection Agency, Washington D.C.: U.S. Government Printing Office, 1993.

Pitt, R., Parmer, K., Clark, S., Field, R. *Potential Groundwater Contamination from Intentional and Nonintentional Stormwater Infiltration,* EPA/600/SR-94/051. United States Environmental Protection Agency, Washington D.C.: U.S. Government Printing Office, 1994.

Plummer, D.H., Owens, N.J.P., Herbert, R.A.. Bacteria–particle interactions in turbid estuarine environments. *Continent. Shelf. Res.* 1987, *7*, 1429–1433.

Pommepuy, M., Butin, M., Derrie, A., Gourmelon, M., Colwell, R.R, Cormier, M. Retention of enteropathogenicity by viable but nonculturable *Escherichia coli* exposed to seawater and sunlight. *Appl. Environ. Microbiol*. 1996, *62(12)*, 4621-4626.

Pommepuy, M., Guillaud, J.F., Dupray, E., Derrien, A., LeGuyader, F., Cormier, M. Enteric bacteria survival factors. *Water Sci. Technol.* 1992, *25*, 93–103.

Powell, M., Ebel, E., Walderhaug, M., Kause, J. Dose Response Envelope for Escherichia coli O157:H7. *Quantitative Microbiology*. 2000, 2, 141–163.

Prieto, M.D., Lopez, B., Juanes, J.A., Revilla, J.A., Llorca, J., Delgado-Rodriguez, M. Recreation risks in coastal waters: health risks associated with bathing in sea water. *J. Epidemiol. Comm. Health*. 2001, *55*, 442-447.

Pruss, A. Review of epidemiological studies on health effects from exposure to recreational water. *Int. J. Epidemiol.* 1998, *271*, 1-9.

Rajal, V., Thompson, D., Kildare, B., Tiwari, S., Mcswain, B., Wuertz, S. *Management of Pathogens Associated with Stormwater Discharge: Methodology for Quantitative Molecular Determination of Viruses, Bacteria and Protozoa, Interim Report,* Sacramento, CA: California Department of Transportation, Environmental Division, 2005.

Ram, J.L., Ritchie, R.P., Fang, J., Gonzales, F.S., Selegean, J.P. Sequence based source tracking of *Escherichia coli* based on genetic diversity of glucuronidase. *J. Environ. Qual.* 2004, *33*, 1024-1032.

Rand, B., Melton, I.E. Particle interactions in aqueous kaolinite suspensions I. Effect of pH and electrolyte upon the mode of particle interaction in homoionic sodium kaolinite suspensions. *J. Coll. Inter. Sci.* 1977, *602*, 308-320.

Rao, S.S., Dutka, B.J., Taylor, C.M. *Ecotoxicological Implications of Fluvial Suspended Particles*, Boca Raton, FL.: Lewis Publishers, 1993.

Rao, V.C., Seidel, K.M., Goyal, S.M., Metcalf , T.G., Melnick, J.L. Isolation of enteroviruses from water, suspended solids, and sediments from Galveston Bay: Survival of poliovirus and rotavirus adsorbed to sediments. *Appl. Environ. Microbiol.* 1984, *48(2)*, 404-409

Rasmus, J. Weldon, K. Moonlight Beach urban runoff treatment facility. *Stormwater- J. Surf. Water Prof.* 2003, *May/June*, 12-22.

Reddy, K.R., Khaleel, R., Overcash, M.R. Behavior and transport of microbial pathogens and indicator organisms in soils treated with organic wastes. *J. Environ. Qual.* 1981, *103*, 255-265.

Refsum, T., Heir, E., Vardund, G.K.T., Holstand, G. Molecular epidemiology of *Salmonella enterica* serovar Typhimurium isolates determined by pulsed-field gel electrophoresis: comparison of isolates from avian wildlife, domestic animals, and the environment in Norway. *Appl. Eniviron. Microbiol.* 2002, *68*, 5600-5606.

Regli, S., Rose, J.B., Haas, C.N., Gerba, C.P. Modeling the risk from *Giardia* and viruses in drinking-water. *J. AWWA*. 1991, *83*(11), 76–84.

Rendtorff, R.C., The Experimental Transmission of *Giardia lamblia* Among Volunteer Subjects. Waterborne Transmission of *Giardia*sis, Proceedings of a Symposium, Cincinnati, OH, 1978.

Reynolds, K.A., Gerba, C.P., Pepper, I.L. Detection of infectious enteroviruses by an integrated cell culture-PCR procedure. *Appl. Environ. Microbiol* 1996, *62*, 1427.

Reynolds, K.A., Gerba, C.P., Pepper, I.L. Rapid PCR-based monitoring of infectious entervirus in drinking water. *Water Sci. and Technology* 1997, *35*, 423-427.

Reynolds, K.A., Roll, K., Fujioka, R.S. Incidence of enteroviruses in Mamala Bay, Hawaii using cell culture and direct polymerase chain reaction methodologies. *Can. J. Microbiol* 1998**,** *44*, 598-604.

Rhodes, M.W., Kator, H.I. Survival of *Escherichia coli* and *Salomonella spp*. in estuarine environments. *Appl. Environ. Microbiol.* 1988, *5412*, 2902-2907.

Rhodes, M.W., Kator, H.I. Effects of sunlight and autochthonous microbiota on *Escherichia coli* survival in an estuarine environment. *Curr. Microbiol*. 1990, 21: 65-73.

Rittenberg, S.C., Mittwer, T., Ivler, D. Coliform bacteria in sediments around three Mine sewage outfalls. *Limnol. Ocean.* 1958, *31*, 101-108.

Ritter, J.K., Carruthers, E.A., Carson, C.A., Ellender, R.D., Harwood, V.J., Kingsley, K., Nakatsu, C.H., Sadowsky, M.J., Shear, B., West, B., Whitlock, J.E., Wiggins, B., Wilbur, J.D. Assessment of statistical methods used in library-based approaches to microbial source tracking. *J. Water Health*. 2003, *01(4)*, 209-233.

Robertson, L.J., Campbell, A.T., Smith, H.V. Survival of *Cryptosporidium parvum* oocysts under various environmental pressures. *Appl. Environ. Microbiol*. 1992, *58(11)*, 3494-3500.

Rochelle, P.A., Ferguson, D.M., Handojo, T.J., De Leon, R., Stewart, M.H., Wolfe, R.L. An assay combining cell culture with reverse transcriptase PCR to detect and determine the infectivity of waterborne *Cryptosporidium parvum*. *Appl. Environ. Microbiol*. 1997 *63*, 2029-2037.

Rollins, D.M., Colwell, R.R. Viable but nonculturable state of *Campylobacter jejuni* and its role in survival in the natural aquatic environment. *Appl. Environ. Microbiol.* 1986, *523*, 531-538.

Roper, M.M., Marshall, K.C. Effects of a clay mineral on microbial predation and parasitism of *Escherichia coli*. *Micro. Ecol.* 1978, *4*, 279-289.

Roper, M.M., Marshall, K.C. Effects of Salinity on Sedimentation and of particulates on survival of bacteria in estuarine habitats. *Geomicrobiology Journal* 1979, *1*(2), 103-116.

Rose, J.B., Mullinax, R.L., Singh, S.H., Yates, M.V., Gerba C.P. Occurrence of rotaviruses and enteroviruses in recreational waters of Oak Creek, Arizona. *Wat. Res.* 1987, *2111*, 1375-1381.

Rose, J.B., Haas, C.N., Regli, S. Risk assessment and control of waterborne *Giardia*sis. *Amer. J. of Pub.Hlth*., 1991, 81(6)709-713.

Rylander, R., Andersson, K., Belin, L., Berglund, G., Bergstrom, R., Hanson, L.A., Lundholm, M., Mattsby, I. Sewage worker's syndrome. *Lancet*. 1976, *2*, 478–479.

Sakamoto, G., Cairns, W.L. UV Dose-response Determination: Collimated Beam vs. Pilot Studies. In *Proceedings of Hawaii Water Environment Association Conference*, Honolulu, HI. 1997.

Santa Clara Valley Urban Runoff Pollution Prevention Program (SCVURPPP). *Application of Water-Quality Engineering Fundamentals to the Assessment of Stormwater Treatment Devices*. Oakland, CA: SCVURPPP, 2000.

Schaub, S. A Risk Assessment Framework for Waterborne Pathogens and Requirements for Producing a Complete Protocol. *Human and Ecological Risk Assessment*, 2004, *10*, 151-158

Schets, F.M., Engels, G.B., During, M., de Roda Husman, A.M. Detection of infectious *Cryptosporidium* oocysts by cell culture immunofluorescence assay: applicability to environmental samples. *Appl. Environ. Microbiol*. 2005, *71*, 6793-6798.

Schiff, K., Kinney, P. Tracking sources of bacterial contamination in stormwater discharges to Mission Bay, California. *Water Environ. Res.* 2001, *73*, 534-542.

Schillinger, J.E., Gannon, J.J. Bacterial Adsorption and Suspended Particles in Urban Stormwater. *Journal of the Water Pollution Control Federation* 1985, *57*(5), 384-389.

Schroeder, E.D., Stallard, W.M., Thompson, D.E., Loge, F.J., Deshussess, M.A., Cox, H.H. *Management of Pathogens Associated with Storm Drain Discharge.* California Department of Transportation, Division of Environmental Analysis: University of California, Davis, 2002.

Schueler, T.R. *A Current Assessment of Urban Best Management Practices*, Publication Number 87703, Washington, D.C: Metropolitan Washington Council of Governments, 1992.

Schueler, T.R. *Controlling Urban Runoff: A Practical Manual for Planning and Designing Urban BMPs*. Washington, D.C.: Metropolitan Washington Council of Governments. 1987.

Schwartz, J., Levin, R., Goldstein, R. Drinking water turbidity and gastrointestinal illness in the elderly of Philadelphia. *J. Epidemiol. Comm. Health.* 2000, *54(1)*, 45-51.

Schwartz, J., Levin, R., Knashawn, H. Drinking water turbidity and pediatric hospital use for gastrointestinal illness in Philadelphia. *Epidemiol.* 1997, *8(6)*, 625-620.

Schwartzbrod, J., Stien, J.L., Bouhoum, K., Baleux, B. Impact of Wastewater Treatment on Helminth Eggs. *Water Science and Technology* 1989, *21*(3), 295-297.

Scott, W. Sanitary study of shore bathing water. *Bulletin of Hygiene* 1951, 33, 353.

Scott, T.M., Rose, J.B., Jenkins, T.M., Farrah, S.R., Lukasik, J. Microbial source tracking: current methodology and future directions. *Appl. Environ. Microbiol.* 2002, *68*, 5796-5803.

Searcy, K.E., Packman, A.I., Atwill, E., Harter, T. Association of *Cryptosporidium parvum* with suspended particles: impact on oocyst sedimentation. *Appl. Environ. Microbiol.* 2005, *712*, 1072-1078.

Seattle Metro and Washington Department of Ecology (SMWE). *Biofiltration Swale Performance: Recommendations and Design Considerations*, Publication No. 657, Water Pollution Control Department: Seattle, WA. 1992.

Selvakumar, A., Borst, M. Variation of microorganism concentrations in urban stormwater runoff with land use seasons. *J. Water Health.* 2006, *4(1)*, 109-24.

Serrano, E., Moreno, B., Solaun, M., Aurrekoetxea, J.J., Ibarluzea J. The influence of environmental factors on microbial indicators of coastal water pollution. *Wat. Sci. Tech.* 1998, *3812*, 195-199.

Seurinck, S., Verstraete, W., Siciliano, S.D. Use of 16s-23s RNA intergenci spacer region PCR and repetitive extragenic palindromic PCR analyses of *Escherichia coli* isolates to identify nonpoint fecal sources. *Appl. Environ. Microbiol.* 2002, *69*, 4942-4950.

Seyfried, P.L, Tobin, R.S., Brown, N.E., Ness, P.F. A prospective study of swimming-related illness. I. Swimming-associated health risk. *Am. J. Public Health* 1985, *75*, 1068-1070.

Shibataa, T., Solo-Gabriele, H.M., Fleming, L.E., Elmir, S. Monitoring marine recreational water quality using multiple microbial indicators in an urban tropical environment. *Wat. Res.* 2004, *38*, 3119–3131

Simpson, J., Santo Domingo, J.W., Reasoner, D.J. Microbial source tracking: state of the science. *Environ. Sci. Tech.* 2002, *36*, 5279-5288.

Sinton, L.W., Davies-Colley R.J., Bell, R.G. Inactivation of enterococci and fecal coliforms from sewage and meatworks effluents in seawater chambers. *Appl. Environ. Microbiol.* 1994, *606*, 2040-2048.

Skinhoj, P., Hollinger, F.B., Hovind-Hougen, K., Lous, P. Infectious liver diseases in three groups of Copenhagen workers: correlation of hepatitis A infection to sewage exposure. *Arch. Environ. Health.* 1981, 36, 139–43.

Sleath, J.F.A. *Sea Bed Mechanics*, New York: John Wiley & Sons, 1984, 335 p.

Smalla, K., Wachtendorf, U., Heuer, H., Liu, W., Forney, L. Analysis of BIOLOG GN substrate utilization patterns by microbial communities. *Appl. Environ. Microbiol.* 1999, *64*, 1220-1225.

Smith, J.J., Perdek, J.M. Assessment and management of watershed microbial contaminants. *Crit. Rev. Environ. Sci. Tech.* 2004, *34*, 109-139.

Smith, J.H., Wickham, J.D., Norton, D., Wade, T.G., Jones., K.B. Utilization of landscape indicators to model potential pathogen impaired waters. *J. Am. Water Res. Assoc.* 2001 *37(4)*, 805-814.

Sobsey, M., Battigelli, D., Handzel, T., Schwab, K. *Male-Specific Coliphages as Indicators of Viral Contamination of Drinking Water*, Denver, CO: American Water Works Association Research Foundation, 1995, 150 p,

Solic, M., Krstulovic, N. Separate and combined effects of solar radiation, temperature, salinity, and pH on the survival of faecal coliforms in seawater. *Mar. Poll. Bull.* 1992, *24(8)*, 411-416.

Soller, J.A., Eisenberg, J.N.S., Degeorge, J., Cooper, R., Tchobanoglous, G., Olivieri A.W.A. Public health evaluation of recreational water impairment. *J. Water Health.* 2006, *41*, 1-23.

Soller, J.A., Olivieri, AW., Eisenberg, J.N.S., Sakaji, R., Danielson, R. *Evaluation of Microbial Risk Assessment Techniques and Applications*, Project 00-PUM-3. Alexandria, VA: Water Environment Research Foundation Report, 2004a.

Soller, J., Stephenson, J., Olivieri, K., Downing, J., Olivieri, A. *Evaluation of First Flush Pollutant Loading and Implications for Water Resources and Urban Runoff Management.* Proceedings of the AWWA 2004 Water Sources Conference. Prepared for the Santa Clara Valley Urban Runoff Pollution Prevention Program by EOA, Inc. and the City of San Jose, CA. 2004b.

Soller, J., Turbow, D., Olivieri, A. *Characterization of the Potential Adverse Human Health Effects Associated with Combined Sewer Overflows and Sanitary Sewer Overflows*. Report for the Association of Metropolitan Sewerage Agencies, EOA Inc: Oakland, CA, 2003.

Solo-Gabriele, H.M., Wolfert, M.A., Desmarais, T.M., Palmer, C.J.. Sources of *Esherichia coli* in a coastal subtropical environment. *Appl. Environ. Microbiol.* 2000, *661*, 230-237.

Southeastern Wisconsin Regional Planning Commission (SWRPC). *Costs of Urban Nonpoint Source Water Pollution Control Measures*, Waukesha, WI: SWRPC, 1991.

Steele, M., Mcnab, B., Fruhner, L., Degrandis, S., Woodward, D., Odumeru, J.A. Epidemiological typing of *Campylobacter* isolates from meat processing plants by pulsed-field gel electrophoresis, fatty acid profile typing, serotyping, and biotyping. *Appl. Environ. Microbiol.* 1998, *64*, 2346-2349.

Steets, B.M., Holden, P.A. A Mechanistic model of runoff associated fecal coliform fate and transport through a coastal lagoon. *Wat. Res.* 2003, *37*, 589-608.

Stenstrom, M. Correspondance from Dr. Michael Strenstorm (UCLA Department of Civil Engineering) to Mr. Paul Corn of United Pumping Services (manufacturer of Drain Pac Filters) detailing results of lab tests performed on Drain Pac Filters. September 25, 1998.

Stern, M.J., Ames, G.F.-L., Smith, N.H., Robinson, E.C., Higgins, C.F. Repetitive extragenic palindromic sequences: a major component of the bacterial genome. *Cell.* 1984, *37*, 1015-1026.

Stevenson, A. Studies of bathing water quality and health. *Am. J. Public Health Assoc.* 1953, *43*, 529.

Stewart, J.R., Ellender, R.D., Gooch, J., Jiang, S., Myoda, S.P., Weisberg, S.B. Recommendations for microbial source tracking: Lessons learned from a methods comparison study. *J. Wat. Health.* 2003, *01(4)*, 228-231.

Stoeckel, D.M., Mathes, M.V., Hyer, K.E., Hagedorn, C., Kator, H., Lukasik, J., O'Brien, T.L., Fenger, T.W., Samadpour, M., Strickler, K.M., Wiggins, B. Comparison of seven protocols to identify fecal contamination sources using *Escherichia coli*. *Environ. Sci. Tech.* 2004, *38*, 6109-6117.

Stokes, W.R. Report of the committee on bathing places. *American Journal of Public Health* 1927a, 17, 121.

Stokes, W.R. A search for pathogenic bacteria in swimming pools. *American Journal of Public Health* 1927b, 17, 334.

Stott, R., May, E., Matsushita, E., Warren, A. Protozoan predation as a mechanism for the removal of *Cryptosporidium* oocysts from wastewaters in constructed wetlands. *Wat. Sci. Tech.* 2001, *44*, 191-198.

Stott, R., May, E. Warren, A. Predation of *Cryptosporidium* oocysts by protozoa and rotifers: Implications for water quality and public health. *Wat. Sci. Tech.* 2003, *473*, 77-83.

Stotzky, G. Influence of clay minerals on microorganisms. III. Effect of particle size, cation exchange capacity, and surface area on bacteria. *Can. J. Microbiol.* 1966, *12*, 1235-1246.

Stotzky, G., Rem, L.T. Influence of clay minerals on microoganisms. I. Montmorillonite and kaolinite on bacteria. *Can. J. Microbiol.* 1966, *12*, 547-563.

Strauch, G., Bittkau, A., Schlosser, D., Petzoldt, H., Sbjeschnid, A. Assessing the pollution of water protection areas and river banks along the Elbe and Mulde Rivers after the flood event in August 2002. *Acta Hydrochim. Hydrobiol.* 2005, *33(5)*, 418 – 429.

Strecker, E., Quigley, M., Urbanas, B.R., Jones, J.E., Clary, J.K. Determining urban storm water BMP effectiveness. *J. Wat. Res. Plan. Manag.* 2001, 144-149.

Streeter, H.W. Bacterial Quality Objectives for the Ohio River: A Guide for the Evaluation of Sanitary Condition of Waters Used for Potable Supplies and Recreational Uses. Cincinnati, Ohio: Ohio River Valley Water Sanitation Commission, 1951.

Suptel, E.A. Pathenogenesis of experimental Coxsackie virus infection. *Arch. Virol.* 1963, *7*, 61–66.

Tate III, R.L. Cultural and environmental factors affecting the longevity of *Escherichia coli* in histosols. *Appl. Environ. Microbiol.* 1978, *355*, 925-929.

Tate, K.W., Atwill, E., George, M., McDougald, N.K., Larsen, R.E. *Cryptosporidium parvum* transport from cattle fecal deposits on California rangelands. *J. Range Manage.* 2000, *533*, 295-299.

Teunis, P.F.M., Van der Heijden, O.G., Van der Giessen, J.W.B., Havelaar, A.H. The dose-response relation in human volunteers for gastro-intestinal pathogens. RIVM Report 1996.

Teunis, P.F.M., Havelaar, A. *Cryptosporidium* in Drinking Water: Evaluation of the ISLI/RSI Quantitative Risk Assessment Framework. 1999.

Thomas, C., Hil, D.J., Mabey, M. Evaluation of the effect of temperature and nutrients on the survival of *Campylobacter spp.* in water microcosms. *J. Appl. Microbiol.* 1999, *86*, 1024-1032.

Thompson, D.E. Rajal, V.B., De Batz, S., Wuertz, S. Detection of *Salmonella* spp. in water using magnetic capture hybridization combined with PCR or real-time PCR. *J Water Health* 2006, 4, 67-75.

U.S. Environmental Protection Agency (U.S. EPA). Quality Criteria for Water. Washington, D.C. 1976.

U.S. Environmental Protection Agency (U.S. EPA). *Results of the Nationwide Urban Runoff Program, Volume 1 Final Report.* EPA 832/R-83/112. U.S. EPA, Washington, D.C.: U.S. Government Printing Office, 1983.

U.S. Environmental Protection Agency (U.S. EPA). *Ambient Water Quality Criteria for Bacteria*, EPA440/5-84/002, U.S. EPA, Washington, D.C.: U.S. Government Printing Office, 1986.

U.S. Environmental Protection Agency (U.S. EPA). *Framework for Ecological Risk Assessment. Risk Assessment Forum*, EPA/630/R-92/001, Washington, D.C.: U.S. Government Printing Office, 1992a.

U.S. Environmental Protection Agency (U.S. EPA). *NPDES Stormwater Sampling Guidance Document.* EPA833/8/92/001. U.S. EPA, Washington, D.C.: U.S. Government Printing Office, 1992b.

U.S. Environmental Protection Agency (U.S. EPA). *Storm Water Technology Fact Sheets*, United States Environmental Protection Agency, Washington, D.C.: U.S. Government Printing Office, 1999a.

U.S. Environmental Protection Agency (U.S. EPA). *Preliminary Data Summary of Urban Stormwater Best Management Practices*, EPA-821-R-99-012, Washington, D.C.: U.S. Government Printing Office, 1999b.

U.S. Environmental Protection Agency (U.S. EPA). *Summary of the August 14-15, 2002, Experts Workshop on Public Health Impacts of Sewer Overflows*, EPA833/R-02/02. U.S. EPA, Washington, D.C.: U.S. Government Printing Office, 2002a.

U.S. Environmental Protection Agency (U.S. EPA). *Considerations in the Design of Treatment Best Management Practices BMPs to Improve Water Quality*. EPA 600/R-03/103. U.S. EPA, Washington, D.C.: U.S. Government Printing Office, 2002b.

U.S. Environmental Protection Agency (U.S. EPA). *The Use of Best Management Practices (BMPs) in Urban Watersheds,* EPA/600/R-04/184, Office of Research and Development, Washington, D.C.: U.S. Government Printing Office, 2004.

United States Geological Survey (USGS). National Land Cover Data. http://eros.usgs.gov/. 1992

United States Geological Survey (USGS). *Fecal Coliform Bacteria Removal Efficiency for Stormwater BMPs in the Virgin Islands*. USGS. #2003VI9B. USGS, Washington, D.C.: U.S. Government Printing Office, 2004.

U.S. Treasury Department. Bacterial standard for drinking water. Public Health Report 1914, 29, 2959-2966.

Van Asperen, I.A., Medema, G., Borgdorff, M.W., Sprenger, M.J., Havelaar, A.H. Risk of gastroenteritis among triathletes in relation to faecal pollution of fresh waters, *Int. J. Epidemiol.* 1998, *27*, 309-315.

Van Donsel, D.J., Geldreich, E.E., Clarke, N.A. Seasonal variations in survival of indicator bacteria in soil and their contribution to storm-water pollution. *Appl. Microbiol.* 1967, *15(6)*,1362-1370.

Vasconcelos, G.J., Swartz, R.G. Survival of bacteria in seawater using a diffusion chamber apparatus in situ. *Appl. Environ. Microbiol.* 1976, *31(6),* 913-920.

Venczel, L., Brown, S., Frumkin, H., Simmonds-Diaz, J., Deitchman, S., Bell, B.P. Prevalence of hepatitis a virus infection among sewage workers in Georgia. *Am. J. Ind. Med.* 2003, *432*, 172-8.

Versalovic, J., Koeuth, T., Lupski, J.R. Distribution of repetitive DNA sequences in eubacteria and application to fingerprinting of bacterial genomes. *Nucl. Acids Res.* 1991, *19*, 6823-6831.

Verstraete, W. Voets, J.P. Comparative study of *E. coli* survival in two aquatic ecosystems. *Wat. Res.* 1976, *10*, 129-136.

Vinje, J., Oudejans, S.J.G., Stewart, J.R., Sobsey, M.D., Long, S.C. Molecular detection and geotyping of male-specific coliphages by reverse transcription-PCR and reverse line blot hybridization. *Appl. Environ. Microbiol.* 2004, *70*, 5996-6004.

Von Schirnding, Y.E., Kfir, R., Cabelli, V., Franklin, L., Joubert, G. Morbidity among bathers exposed to polluted seawater. A prospective epidemiological study. *S. African Med. J.* 1992, *81(11)*, 543-6.

Vonstille, W.T., Stille, W.T., III, Sharer, R.C. Hepatitis A epidemics from utility sewage in Ocoee, Florida. *Arch. Environ. Health*. 1993, *48(2)*, 120-4

Wade, T.J., Pai, N., Eisenberg, J., Colford, J.M. Do U.S. EPA water quality guidelines for recreational waters prevent gastrointestinal illness? A systematic review and meta-analysis. *Environ. Health Perspect.* 2003, *111*, 1102-1109.

Wade, T.J., Sandhu, S.K., Levy, D., Lee, S., Lechevallier, M.W., Katz, L., Colford Jr., J.M. Did a severe flood in the midwest cause an increase in the incidence of gastrointestinal symptoms? *Am. J. Epidemiol.* 2004, *1594*, 398-405.

Wait, D.A., Sobsey, M.D. Comparative survival of enteric viruses and bacteria in Atlantic Ocean seawater. *Water Sci. Technol.* 2001, *4312*, 139-142.

Ward, R.L., Bernstein, D.L., Young, C.E., Sherwood, J.R., Knowlton, D.R., Schiff, G.M. Human rotavirus studies in volunteers: determination of infectious dose and serological response to infection. *J. Infect. Dis.*1986, *154*(5), 871.

Walker, S.L. The influence of bacterial surface polymers on bacterial adhesion and transport in groundwater environments. *J. Harbin Inst. Tech.* 2005, *12*, 19-26.

Water Environment Research Foundation (WERF). *Proceedings from Microbial Source Tracking Workshop.* San Antonio, TX: Water Environmental Research Foundation, 2005.

Watershed Management Institute (WMI). *Operation, Maintenance, and Management of Stormwater Management Systems*, Report for the U.S. Environmental Protection Agency, Office of Water, D.C.: U.S. Government Printing Office Washington, D.C. 1997.

Wayne County. *Planning and Cost Estimating Criteria for Best Management Practices,* Wayne County, FL, 2001.

Weldon, M., Vanegdom, M.J., Hendricks, K.A., Regner, G., Bell, B.P., Sehulster, L.M. Prevalence of antibody to hepatitis a virus in drinking water workers and wastewater workers in Texas from 1996 to 1997. *J. Occup. Environ. Med.* 2000, *428*, 821-6.

West, T.A., Sutherland, J.W., Bloomfield, J.A. *A Study of the Effectiveness of a Vortechs^{Tm} Stormwater Treatment System for Removal of Total Suspended Solids and Other Pollutants in the Marine Village Watershed, Village of Lake George, New York.* Albany, NY: NYS Department of Environmental Conservation, 2001.

Wheeler, A.L., Hartel, P.G., Godfrey, D.G., Hill, J.L., Segars, W.I. Potential of *Enterococcus faecalis* as a human fecal indicator for microbial source tracking. *J. Environ. Qual.* 2002, *31*, 1286-1293.

Whitman, R.L., Nevers, M.B. Foreshore sand as a source of *Escherichia coli* in nearshore water of a Lake Michigan beach. *Appl. Environ. Microbiol.* 2003, *699*, 5555-5562.

Whitman, R.L., Nevers, M.B., Korinek, G.C., Byappanahalli, M.N. Solar and temporal effects on *Escherichia coli* concentration at a Lake Michigan swimming beach. *Appl. Environ. Microbiol.* 2004, *707*, 4276-4285.

Winfield, M.D., Groisman, E.A. Role of nonhost environments in the lifestyles of *Salmonella* and *Escherichia coli*. Appl. Environ. Microbiol. 2003, *69(7)*, 3687–3694.

Wojtenko, I., Stinson, M.K. *Research Summary. CSO Disinfection Pilot Study: Spring Creek CSO Storage Facility Upgrade,* EPA/600-R-02/077, U.S. Environmental Protection Agency, Office of Research and Development, Cincinnati, OH. 2003.

Wojtenko, I., Stinson, M.K., Field, R. Challenges of combined sewer overflow disinfection by ultraviolet light irradiation. *Crit. Rev. Environ. Sci. Tech.* 2001, *31(3)*, 223-239.

Woodward-Clyde. *Street Sweeping/Storm Inlet Modification Literature Review.* Prepared for the Alameda County Urban Runoff Clean Water Program. December 21, 1994.

World Health Organization (WHO). *Health Based Monitoring of Recreational Waters: The Feasibility of a New Approach (the 'Annapolis Protocol')*, Who/Sde/Wsh/99.1, Geneva, Switzerland, 1999.

Xiao, L., Alderisio, K.' Limor, J., Royer, M., Lal, A. Identification of Species and Sources of *Cryptosporidium* Oocysts in Storm Waters with a Small Subunit RNA-based Diagnostic and Genotyping Tool. *Applied Environmental Microbiology*, 2000, *66*, 5492-98.

Yoder, J.S., Blackburn, B.G., Craun, G.F., Hill, V., Levy, D.A., Chen, N., Lee, S.H., Calderon R.L., Beach, M.J. Surveillance for waterborne-disease outbreaks associated with recreational water--United States, 2001-2002. *MMWR Surveill. Summ.* 2004, *538*, 1-22.

WASTEWATER UTILITY

Alabama
Montgomery Water Works & Sanitary Sewer Board

Alaska
Anchorage Water & Wastewater Utility

Arizona
Glendale, City of, Utilities Department
Mesa, City of
Peoria, City of
Phoenix Water Services Department
Pima County Wastewater Management
Safford, City of

Arkansas
Little Rock Wastewater Utility

California
Calaveras County Water District
Central Contra Costa Sanitary District
Corona, City of
Crestline Sanitation District
Delta Diablo Sanitation District
Dublin San Ramon Services District
East Bay Dischargers Authority
East Bay Municipal Utility District
Eastern Municipal Water District
El Dorado Irrigation District
Fairfield-Suisun Sewer District
Fresno Department of Public Utilities
Inland Empire Utilities Agency
Irvine Ranch Water District
Las Virgenes Municipal Water District
Livermore, City of
Los Angeles, City of
Los Angeles County, Sanitation Districts of
Napa Sanitation District
Orange County Sanitation District
Palo Alto, City of
Riverside, City of
Sacramento Regional County Sanitation District
San Diego Metropolitan Wastewater Department, City of
San Francisco, City & County of
San Jose, City of
Santa Barbara, City of
Santa Cruz, City of
Santa Rosa, City of
South Bayside System Authority
South Coast Water District
South Orange County Wastewater Authority
Stege Sanitary District
Sunnyvale, City of
Union Sanitary District
West Valley Sanitation District

Colorado
Aurora, City of
Boulder, City of
Greeley, City of
Littleton/Englewood Water Pollution Control Plant
Metro Wastewater Reclamation District, Denver

Connecticut
Greater New Haven WPCA

District of Columbia
District of Columbia Water & Sewer Authority

Florida
Broward, County of
Fort Lauderdale, City of
Miami-Dade Water & Sewer Authority
Orange County Utilities Department
Reedy Creek Improvement District
Seminole County Environmental Services
St. Petersburg, City of
Tallahassee, City of
Tampa, City of
Toho Water Authority
West Palm Beach, City of

Georgia
Atlanta Department of Watershed Management
Augusta, City of
Clayton County Water Authority
Cobb County Water System
Columbus Water Works
Fulton County
Gwinnett County Department of Public Utilities
Savannah, City of

Hawaii
Honolulu, City & County of

Idaho
Boise, City of

Illinois
American Bottoms Wastewater Treatment Plant
Greater Peoria Sanitary District
Kankakee River Metropolitan Agency
Metropolitan Water Reclamation District of Greater Chicago
Wheaton Sanitary District

Iowa
Ames, City of
Cedar Rapids Wastewater Facility
Des Moines, City of
Iowa City

Kansas
Johnson County Unified Wastewater Districts
Unified Government of Wyandotte County/ Kansas City, City of

Kentucky
Louisville & Jefferson County Metropolitan Sewer District
Sanitation District No. 1

Louisiana
Sewerage & Water Board of New Orleans

Maine
Bangor, City of
Portland Water District

Maryland
Anne Arundel County Bureau of Utility Operations
Howard County Department of Public Works
Washington Suburban Sanitary Commission

Massachusetts
Boston Water & Sewer Commission
Upper Blackstone Water Pollution Abatement District

Michigan
Ann Arbor, City of
Detroit, City of
Holland Board of Public Works
Saginaw, City of
Wayne County Department of Environment
Wyoming, City of

Minnesota
Rochester, City of
Western Lake Superior Sanitary District

Missouri
Independence, City of
Kansas City Missouri Water Services Department
Little Blue Valley Sewer District
Metropolitan St. Louis Sewer District

Nebraska
Lincoln Wastewater System

Nevada
Henderson, City of
Reno, City of

New Jersey
Bergen County Utilities Authority
Ocean, County of
Passaic Valley Sewerage Commissioners

New York
New York City Department of Environmental Protection

North Carolina
Charlotte/Mecklenburg Utilities
Durham, City of
Metropolitan Sewerage District of Buncombe County
Orange Water & Sewer Authority

Ohio
Akron, City of
Butler County Department of Environmental Services
Columbus, City of
Metropolitan Sewer District of Greater Cincinnati
Northeast Ohio Regional Sewer District
Summit, County of

Oklahoma
Oklahoma City Water & Wastewater Utility Department
Tulsa, City of

Oregon
Clean Water Services
Eugene, City of
Gresham, City of
Portland, City of
Water Environment Services

Pennsylvania
Philadelphia, City of
University Area Joint Authority

South Carolina
Charleston Water System
Mount Pleasant Waterworks & Sewer Commission
Spartanburg Water

Tennessee
Cleveland, City of
Knoxville Utilities Board
Murfreesboro Water & Sewer Department
Nashville Metro Water Services

Texas
Austin, City of
Dallas Water Utilities
Denton, City of
El Paso Water Utilities
Fort Worth, City of
Houston, City of
San Antonio Water System
Trinity River Authority

Utah
Salt Lake City Corporation

Virginia
Alexandria Sanitation Authority
Arlington, County of

WERF SUBSCRIBERS

Fairfax County
Hampton Roads Sanitation District
Henrico, County of
Hopewell Regional Wastewater Treatment Facility
Loudoun County Sanitation Authority
Lynchburg Regional WWTP
Prince William County Service Authority
Richmond, City of
Rivanna Water & Sewer Authority

Washington
Everett, City of
King County Department of Natural Resources
Seattle Public Utilities
Sunnyside, Port of
Yakima, City of

Wisconsin
Green Bay Metro Sewerage District
Kenosha Water Utility
Madison Metropolitan Sewerage District
Milwaukee Metropolitan Sewerage District
Racine, City of
Sheboygan Regional Wastewater Treatment
Wausau Water Works

Australia
South Australian Water Corporation
Sydney Water Corporation
Water Corporation of Western Australia

Canada
Regina, City of, Saskatchewan
Toronto, City of, Ontario
Winnipeg, City of, Manitoba

New Zealand
Watercare Services Limited

United Kingdom
Yorkshire Water Services Limited

STORMWATER UTILITY

California
Fresno Metropolitan Flood Control District
Los Angeles, City of, Department of Public Works
Monterey, City of
Sacramento, County of
San Francisco, City & County of
Santa Rosa, City of
Sunnyvale, City of

Colorado
Aurora, City of
Boulder, City of

Georgia
Griffin, City of

Iowa
Cedar Rapids Wastewater Facility
Des Moines, City of

Kansas
Overland Park, City of

Kentucky
Louisville & Jefferson County Metropolitan Sewer District

Maine
Portland Water District

North Carolina
Charlotte, City of, Stormwater Services

Pennsylvania
Philadelphia, City of

Tennessee
Chattanooga Stormwater Management

Texas
Harris County Flood Control District, Texas

Washington
Bellevue Utilities Department
Seattle Public Utilities

STATE

Arkansas Department of Environmental Quality
Connecticut Department of Environmental Protection
Kansas Department of Health & Environment
Kentucky Department of Environmental Protection
Ohio River Valley Sanitation Commission
Urban Drainage & Flood Control District, CO

CORPORATE

ADS Environmental Services
Advanced Data Mining International
Alan Plummer & Associates
Alden Research Laboratory Inc.
Alpine Technology Inc.
Aqua-Aerobic Systems Inc.
Aquateam–Norwegian Water Technology Centre A/S
ARCADIS
Associated Engineering
Black & Veatch
Boyle Engineering Corporation
Brown & Caldwell
Burgess & Niple, Ltd.
Burns & McDonnell
CABE Associates Inc.
The Cadmus Group
Camp Dresser & McKee Inc.
Carollo Engineers Inc.
Carpenter Environmental Associates Inc.
CDS Technologies Inc.
CET Engineering Services
Chemtrac Systems Inc.
CH2M HILL
Conestoga-Rovers & Associates
CONTECH Stormwater Solutions
D&B/Guarino Engineers, LLC
Damon S. Williams Associates, LLC
Earth Tech Inc.
Ecovation
EMA Inc.
Environ/The ADVENT Group, Inc.
Fay, Spofford, & Thorndike Inc.
Freese & Nichols Inc.
ftn Associates Inc.
Gannett Fleming Inc.
Garden & Associates, Ltd.
Geosyntec Consultants
GHD
Golder Associates Ltd.
Greeley and Hansen LLC
Hazen & Sawyer, P.C.
HDR Engineering Inc.
HNTB Corporation
Hydromantis Inc.
HydroQual Inc.
Infilco Degremont Inc.
Jacobson Helgoth Consultants Inc.
Jason Consultants LLC Inc.
Jordan, Jones, & Goulding Inc.
KCI Technologies Inc.
Kelly & Weaver, P.C.
Kennedy/Jenks Consultants
KMK Consultants
Komline Sanderson Engineering Corporation
Limno-Tech Inc.
Lombardo Associates Inc.
The Low Impact Development Center Inc.
Malcolm Pirnie Inc.
Material Matters
McKim & Creed
Metcalf & Eddy Inc.
Monteco Corporation
MPR Engineering Corporation, Inc.
MWH
NewFields Water Resources, LLC
O'Brien & Gere Engineers Inc.
Odor & Corrosion Technology Consultants Inc.
Oscar Larson & Associates
PA Government Services Inc.
Parametrix Inc.
Parsons
Post, Buckley, Schuh & Jernigan
Praxair, Inc.
RMC Water & Environment
Ross & Associates Ltd.
Rothberg, Tamburini & Windsor, Inc.
SAIC
Savin Engineers
Siemens Water Technologies
Stantec Consulting Inc.
Stearns & Wheler, LLC
Stone Environmental Inc.
Stratus Consulting Inc.
Synagro Technologies Inc.
Tetra Tech Inc.
Trojan Technologies Inc.
Trussell Technologies, Inc.
URS Corporation
Wade-Trim Inc.
Westin Engineering Inc.
Weston Solutions Inc.
Woodard & Curran
Zenon Environmental Inc.
Zoeller Pump Company

INDUSTRY

American Electric Power
American Water
ChevronTexaco Energy Research & Technology Company
The Coca-Cola Company
Dow Chemical Company
DuPont Company
Eastman Chemical Company
Eli Lilly & Company
Merck & Company Inc.
Premier Chemicals LLC
Procter & Gamble Company
Thames Water Utilities
Severn Trent Services Inc.
Suez Environnment
United Water Services LLC

Note: List as of 2/1/07

IWA Publishing's authorised EU representative for General Product Safety Regulations is Diane D'Arras, 15 rue Duret, 75116 Paris, France, e-mail: safety@iwap.co.uk.

Printed and bound by CPI Group (UK) Ltd, Croydon, CR0 4YY
12/05/2026
02108893-0004